SAS COMBAT HANDBOOK

SAS COMBAT HANDBOOK

Barry Davies, BEM

Skyhorse Publishing

Skyhorse Publishing books may be purchased in bulk at special discounts for sales promotion, corporate gifts, fund-raising, or educational purposes. Special editions can also be created to specifications. For details, contact the Special Sales Department, Skyhorse Publishing, 307 West 36th Street, 11th Floor, New York, NY 10018 or info@skyhorsepublishing.com.

Skyhorse® and Skyhorse Publishing® are registered trademarks of Skyhorse Publishing, Inc.®, a Delaware corporation.

Visit our website at www.skyhorsepublishing.com.

10 9 8 7 6 5 4 3 2 1

Library of Congress Cataloging-in-Publication Data is available on file.

Cover photos: Thinkstock

Print ISBN: 978-1-63220-295-6
Ebook ISBN: 978-1-63220-866-8

Printed in China

To "Jock" Logan my best friend while serving in the SAS

TABLE OF CONTENTS

It is a sad thing to say but beneath the soil of most countries on this planet there rests the body of an SAS soldier. On average, three SAS soldiers die every year; on a bad year it can be as high as twenty. The SAS is not a large Regiment, two hundred active men at best, and the death of an individual soldier is a great loss. It is an unfortunate part of SAS life that many soldiers do die, some while in training while others during operations. One only has to take a look in the SAS graveyard at Saint Martin's Church in Hereford or the Clock Tower in Credenhill to see the cost. Here lie (in many cases only named) many young men, good men, men full of life and promise, dead before their time. It is not that the regiment is careless or "Gung-Ho"; on the contrary, it is because these men dared to face foe and adversity head on—they "Dared to Win." The ode below is a fitting verse.

* * *

"It is not the critic who counts, not the one who points out how the strong man stumbled, or how the doer of deeds might have done better. The credit belongs to the man who is actually in the arena, whose face is marred with sweat and dust and blood, who strives valiantly; who knows the enthusiasms, the great devotions, and spends himself in a worthy cause; who, if he wins, knows the triumph of high achievement; and who, if he fails, at least fails while daring greatly, so that his place shall never be with those cold and timid souls who knew neither victory nor defeat."

INTRODUCTION

Writing a book on SAS tactics and operational skills is not as clear cut as one would think, and in truth it would take volumes. Their role within the British Army and Security Services makes instruction on an SAS skill prohibited, and it would be wrong of me to do so. However, it is clear to see how many of these skills came about by highlighting them in the various military operations the SAS have undertaken since their conception in 1941. Therefore, in writing this book, I have selected a range of SAS operations covering most of the Regiment's skills. Some have been widely publicized while others remain untold; none presents a danger to any serving member of the SAS, yet they make for avid reading. With an understanding of the nature of the task confronting the SAS when presented with an operation, it is easier to comprehend the tactics and skills illustrated in the summary. For the most part, the stories within this book come from a direct source where the actual soldiers involved have given their own accounts. Readers might like to see the bibliography at the end of this book and in particular read the SAS War Diary (if you can find one).

The SAS have been asked to do many things, and in the past, most would have made headline news. Nevertheless, recent years have seen a marked decline in the amount of information seeping out of the Regiment. Today secrecy is rigidly enforced. Now, unless operationally required, no SAS soldier is allowed to take a camera with him on operations. All serving Sabre Squadron soldiers (known as Blades) are required to sign a document, which is little more than a vow of silence—for life.

The stories collected within this book have taken place, but the accuracy will always be argued by those who actually took part. The truth can only be told by those who were actually there—even then it is only their version. There is only one truth—the SAS will always be at the cutting edge of warfare, developing its own tactics and operational skills as required.

Some seventy years or more after its inception, news of Special Air Service (SAS) operations still holds a fascination for so many, both at home and abroad. Their daring, their swiftness and bravery are unrivaled in the fight against terrorism. Just two hundred men rigorously selected, highly trained, and with a spirit to dare. They will go, willingly, deep behind enemy lines, take on incredible odds, and risk their lives to rescue others. There have been many books written on the SAS, most telling of individual events in its history. This book aims to describe how the tactics and operational skills help its members win their battles.

In order to explain how SAS tactics and operations work, it is much easier to relate to their role at a given time in their short history. These roles covered operations in the jungle, desert, and cities; but the roles were not dictated by terrain alone and rather by many other factors such as the enemy they faced, the indigenous people, and most importantly, what was expected of the SAS. For example, the London Iranian Embassy siege required tactics far different from those of SAS soldiers operating in Northern Ireland or Afghanistan. To this end I have tried to depict a real operational scenario and illustrate how the SAS adapted their tactics to best meet the operational demands.

For every SAS soldier, the basic war-fighter skills are taught during initial training upon induction into the army. It is only when a soldier volunteers for the SAS and successfully passes selection that he learns a new set of skills. Then, as the years pass and the soldier's assigned to an SAS Squadron or "Cell," he learns the more specialized skills that set the SAS apart from others.

In order to appreciate the actions of the SAS, one must first recognize what makes it so special. "22 Special Air Service" is the designation given to the present SAS regular regiment. The Regiment is now based in Credenhill, Herefordshire, and consists of around two hundred men divided among four Sabre Squadrons: A, B, D, and G. In addition to these, the Regiment also supports the Training Squadron, 18 Signals Regiment, and HQ squadron. There are a number of smaller units such as Operations Capability,

Demolitions, Parachute Section, Boat Section, Army Air Corp Section, and a host of subunits responsible for the daily running and administration of such a large organization. In total the new camp at Credenhill houses almost a thousand service men and women.

Many reading this book will wonder where SAS soldiers come from. What makes an SAS soldier? Are they born killers? Having known many of the men mentioned in this book, I can put your mind at rest: for the most part they are all normal. True some of the SAS soldiers I have known and worked with have told me tales of their background and in some cases they are far from normal.

Take "Nobby" for example. At the age of seven he came home from school to find that his mother and father had moved without telling him. He walked the streets not knowing what to do; eventually he went to live with his aunt. Despite being abandoned by his parents, he entered the ranks of the SAS and remained one of the most stable persons I have ever known.

Then there was Steve. He was nine years old when his mother died. His father began drinking heavily and became a down-and-out. With no home life, Steve took his seven-year-old sister and went to live roughly in the streets of Manchester—it was two years before the authorities caught up with them. A few years later, Steve joined the army and finally made his way to the SAS. Once he was established in Hereford, he sent for his sister who eventually married one of Steve's friends.

For the most part they are just soldiers for whom being in the SAS was the pinnacle—and they will all tell you, selection was a real bitch. Such is the nature of SAS operations that they require a high standard of individual soldiers to carry them out. Year after year they continue to come forward.

The Special Air Service also has a history that equals any other military unit in the world. It earned this reputation by quietly suppressing insurgencies, confronting terrorism head on, and doing the government's bidding to everyone's satisfaction. They would go, not in thousands, but in small groups of men trusting in the belief of their training and comrades. For when men of the right caliber accept a common philosophy based on the individual spirit acting in tune with his brother soldiers to form a whole, then excellence can be the only result.

Today that brotherhood encompasses the Special Forces from Australia and New Zealand, and those of the United States. While the SAS has always had a good working relation with the American Special Forces, the past fifteen years have seen a much stronger bond develop. In recent conflicts such as those in Afghanistan and Iraq, the Americans and the SAS have worked as one operational unit, their missions directed by the Department of Defense.

With such fame, there is a public thirst for any information that surrounds the SAS. To quench this thirst and accommodate this "need," this book encompasses the SAS in its fullest spectrum ever. Retired SAS soldiers have told their personal stories, and thirty-six books about the SAS family have been tabulated and cross-referenced. This has resulted in a huge database on all aspects of the SAS family, from its early start to its present role in Afghanistan.

Chapter 1

SAS FOUNDED

It would be wrong to start this book without providing the reader with a little background on the early history of the SAS. While so much has changed over the years, those original SAS soldiers were the first to adapt tactics and skills to suit their operational requirements. They learned from their mistakes, modified their tactics, and honed their skills. In this aspect little has changed over the years other than the great wealth of military knowledge that has been sharpened and refined. Hence, SAS tactics and operational skills started early in WWII.

By February 1941, British armored forces (known as Layforce) had crossed the Libyan Desert to a point south of Benghazi and cut off the retreating Italians. The resulting Battle of Beda, starting on February 5, inflicted heavy losses. Australian troops captured the major port of Benghazi on February 7, and two days later El Agheila was reached. There the advance stopped. Three days later, on February 12, the first units of the Africa Corps under Rommel arrived in Tripoli. The appearance of Rommel and his Corps in North Africa meant Layforce was split up into three units. One, under Laycock, fought in Crete, another was based around Tobruk, and the third unit (8 Commando) was sent to Syria where it took part in a number of raids on the Cyrenaica coastline.

One of the 8 Commando officers, a young Scots Guards Lieutenant by the name of David Stirling, had joined the Scots Guards supplementary reserve in 1938, but in the two years up to Dunkirk, he had spent much of his time traveling to North America, or when back in Britain sitting in his favorite chair at White's Club. Many considered him to be a wastrel, an aristocrat who was easily bored. It was at White's Club that Stirling had first heard of the Commandos and had promptly enlisted.

Stirling had been involved in a number of unsuccessful large-scale raids on enemy targets which were intended to bolster the defenses of Tobruk and to support the withdrawal

Colonel Sir Archibald David Stirling, founder of the Special Air Service.

from Crete. However, almost all of the commando raids were singularly unsuccessful, largely because they were too ambitious and unwieldy. In the end, Middle East Headquarters (MEHQ) decided that Layforce itself was to be disbanded as the men and materials were more urgently needed elsewhere.

Stirling in the meantime had visited his brother Peter in Cairo, who happened to be Third Secretary at the British Embassy (Stirling had three brothers, Peter, Hugh, and Bill). Stirling soon set himself up at his brother's flat and proceeded to fall back into his old ways, with many a night spent partying or visiting the Scottish Hospital where he had befriended a local nurse. It was around this time that David Stirling met a young Lieutenant, Jock Lewes, in the Officers' mess. Socially, the two men were as different as chalk and cheese, yet they soon discovered they shared a common belief. They were convinced the similarity lay in independent small-sized teams of specially trained men who could operate behind enemy lines. Lewes had been particularly impressed with the German parachute assault on the island of Crete, an assault which had cost the Germans dearly (German airborne forces suffered over seven thousand casualties), but had also won them a victory and with it, Crete. Just before Layforce was disbanded, Lewes had discovered some fifty parachutes, which had been landed at Port Said. These parachutes were awaiting destination for India. Lewes approached Laycock for permission to "borrow" a dozen or so and try them out—Laycock approved. Stirling, having heard of this, managed to get himself involved and preparations were made for a parachute jump.

In June 1941, Jock Lewes and his batman, Roy Davies, made the

Lieutenant John Steel Lewes 'Jock' originated the idea of Special Forces operating behind enemy lines which prompted Stirling's vision of the SAS. Lewes is held by many as the co-founder of the SAS. Seen here on the right of the picture with David Stirling.

first jump from an outdated Vickers Valentia biplane; both made it safely. Stirling and Mick D'Arcy jumped second, but Stirling's parachute snagged on the tail section, ripping a large hole in the silk; consequently he had a hard landing. This landing paralyzed Stirling and also caused blindness for a short time, and thus he was sent to the hospital.

Most history books will tell you that it was here, lying in a hospital bed, that Stirling first conceived the idea of the SAS. This is not true. The truth is, Stirling had been bitten by the idea that he and Jock Lewes shared, and his time in the hospital allowed him to write down his thoughts into a proper proposal. Stirling also knew it would be futile to put his ideas through the normal chain of command—he was not popular with most of the junior General Staff officers. Luckily for Stirling, General Auchinleck had replaced General Wavell as Commander-in-Chief after the unsuccessful relief of Tobruk. Auchinleck was known for his like of devil-may-care soldiers; furthermore, he was a friend of the Stirling family.

Stirling understood the benefits of attacking targets behind enemy lines and disrupting Rommel's supply lines, and better still of taking out his aircraft as the Luftwaffe had control of the skies at the time. He felt sure the raids would have a greater chance of success if executed by small groups of men, thus using the element of surprise to its best advantage. He also knew that a small unit of four or five men could operate more effectively, as they would all depend on each other and not on the overriding authority of rank. Raids would also be far more effective if they took place at night. All these ideas were refined and put to paper. The next step was to get his scheme endorsed.

It is stated that Stirling, despite his injuries, managed to gain access to MEHQ by climbing through a hole in the perimeter fence and reaching the office of General Richie, who at the time was Deputy Chief of Staff. Stirling presented his plan to Richie and the memorandum swiftly reached the desk of General Auchinleck. The idea suited Auchinleck's purpose, for he was planning a new offensive later that same year.

Three days later, Stirling was ordered back to MEHQ. The meeting was brief but positive; Stirling was promoted to captain and given authority to recruit six officers and sixty other ranks. Brigadier Dudley Clarke, the man who had come up with the name "Commando" for Churchill, and at the time was running a deception unit at MEHQ, thus assigned Stirling's new command as "L" Detachment, Special Air Service Brigade. This was a hopeful ruse that would fool the Germans into thinking the British had some form of elite troops in the area. And thus, the SAS was born.

David Stirling was determined to build a unit of dedicated men; men of ability and capable of self-discipline. He is quoted as saying, "We believe, as did the ancient Greeks who originated the word 'Aristocracy,' that every man with the right attitude and talents, regardless of birth and riches, has a capacity in his own lifetime of reaching that status in its true sense. In fact, in our SAS context, an individual soldier might prefer to go on serving as an NCO rather than leave the regiment in order to obtain an Officer's commission. All ranks in the SAS are of 'one company,' in which a sense of class is both alien and ludicrous." This ethos remains within the SAS family to the present day.

If David Stirling was the creator of the SAS, then Jock Lewes was its heart. In Stirling's own words, "he was indispensable and I valued him more than I had ever originally appreciated." David Stirling said of Lewes: "Jock could far more genuinely claim to be founder of the SAS than I." If there is a reason why Jock Lewes did not receive full credit for his contribution to the forming of the SAS, it was his early death in December 1941, shortly after the SAS was first formed.

The second man Stirling really wanted was Paddy Mayne. They had known each other during their service in Layforce where Mayne had undertaken one of the few successful operations. The nickname "Paddy" came with his Irish ancestry, and before the war he had been a solicitor and was well known for his accomplishments in the world of sport. In battle he possessed qualities of leadership which set him apart from most men, and a

Lieutenant-Colonel Robert Blair "Paddy" Mayne. All-around athlete and the one man who could get the job done. Mayne's contribution to the SAS was exceptional.

reputation built on his personal bravery, which at times was characterized as reckless and wild. Mayne was to prove that he would be the "fighter," the man who would happily go into battle—and the man other men would happily follow into battle, to make the SAS a success.

When the SAS first arrived at Kabrit, the camp area designated to them, they were greeted by a small board stuck in the sand which read, L DETACHMENT SAS. Kabrit lay 90 miles east of Cairo on the edge of the Great Bitter Lake. There was nothing but sand. No buildings, no tents, no mess hall—nothing. Worse still the new "Brigade" had no weapons or supplies. As there was no parent unit, they had no one to call on for support. Stirling reported directly to General Auchinleck, and while this offered no support in the way of logistics, it was the first move in maintaining independence for the SAS.

The men of the unit begged, borrowed, or stole whatever they needed. Under cover of darkness they stole tents and equipment from the nearby New Zealanders' camp, many of whom were away training in the desert at the time. They raided the Royal Engineers and stole cement and other building material. One of the best scroungers was a Londoner called Kaufman, who reportedly stole enough material from the RAF to construct a proper canteen. Kaufman raided and falsely requisitioned rations and stores which transformed the SAS camp into one of the best bases in the area. Kaufman soon realized that he was not cut out to be an SAS soldier, but he remained with the unit as a store man.

Lewes started the training based on what he expected the challenges to be like when operating deep behind enemy lines. Initially the men did a great deal of hiking through the desert carrying a full load of around 50lbs on their backs. During such hikes they were limited as to the amount of food and water they could carry. When in camp there was instruction on weapons, including British, German, and Italian. While hand grenades were available, it was found that they were not capable of destroying an aircraft with any consistency, and thus the "Lewes Bomb" was introduced.

Author's Note: The bomb was designed by Lieutenant "Jock" Lewes, with the express purpose of destroying Second World War aircraft. The bomb was basic, a mixture of plastic explosive (TNT 808) thermite and a flammable fuel, normally diesel oil. This was attached to the aircraft where the wing met with the fuselage, and normally always on the right wing. A time pencil and a No. 27 detonator initiated the bomb. It was later found that the time-pencil was a glass tube with a spring loaded striker held in place by a copper wire. The top of the tube held a small glass vial of acid, which, when crushed, released the acid and burnt through the wire—the thicker the wire, the longer the delay. These proved very unreliable, and so other methods such as release and pressure switches were used. When completed, the whole bomb was put in a small sock-like bag and covered with a sticky compound so it would adhere to the aircraft.

The SAS also got busy with parachute training, which included jumping backwards off the back of a truck traveling at 30 mph. After several injuries, this method was abandoned and a proper jump training facility was to be designed and built by the nearby engineers. The first structure was a tower for parachute jumps and landing training.

However, despite the amount of ground training, parachuting and parachutes were relatively new and thus, many of the safety checks and procedures had yet to be realized and put into practice. On October 16, 1941, the first practice jump for the SAS took place. The first stick of ten men climbed into a Bristol Bombay aircraft of 216 Squadron which took off and settled at a height of around 900 feet. Several men jumped and landed without incident, but when Ken Warburton, a twenty-one-year-old, jumped, his parachute failed to open and he plummeted to his death. He was followed by Joe Duffy, Warburton's best friend. Records show Duffy had suspicions that something was wrong and queried the jumpmaster sergeant. However, the humiliation of refusing to jump or RTU (Returned to Unit) overpowered his unease and he leapt. Again the parachute did not open and Duffy hit the ground, reportedly quite close to his friend.

Author's Note: It has never been explained who invented the term Returned to Unit (RTU), probably Jock Lewes, but it appears that the discipline was in place from the very outset of the SAS. If a soldier either fails selection or is later expelled from the Regiment for misbehavior or a failing in standards, he is returned to his original unit. The term RTU can galvanize the SAS soldier to do things he would not normally do, even at the risk of personal injury. It is an ethos that exists to this day and applies to both officers and other ranks alike.

The famous SAS winged dagger.

It was around this time that the famous insignia of the SAS materialized. There have been many a discussion on what the emblem depicts; some say it's the flaming sword of Excalibur, while others claim it's a winged dagger. It was Bob Tait who designed the cap badge, and the motto was down to David Stirling who, after listening to many an idea such as "Strike and Destroy," came up with "Who Dares Wins."

By November 1941, L Detachment of the SAS was ready to carry out its first operation code-named Operation "Squatter." The aim of a new offensive Auchinleck, code-named Operation "Crusader" was to retake Cyrenaica and secure the Libyan airfields which at the time were in enemy hands. If this could be achieved then shipping supplies to Malta could be increased; additionally it would open up Sicily for raids—the stepping stone to Italy.

On the night of November 16–17, a force of fifty-five men was divided into five aircraft provided by 216 Squadron and parachuted into the desert behind enemy lines. David Stirling, "Paddy" Mayne, Eoin McGonigal, "Jock" Lewes, and Lieutenant Bonnington commanded this force respectively. The raids would take place against the airfields in the area of Tmimi and Gazala. Rommel was also intending to advance and had plans of his own, as it was widely believed that the German Luftwaffe had received reinforcements in the form of the new Messerschmitt 109s. The parachute entry would drop the SAS twelve miles from the target. Once Operation Squatter was completed they would all rendezvous three miles southeast of the Gadd-el-Ahmar crossroads where the Long Range Desert Group (LRDG) would be waiting to ferry them safely back to Kabrit.

The five Bristol Bombay aircraft took off on schedule. Although the night of take-off was clear and still, as the aircraft proceeded towards their drops zones, the weather quickly changed. Thick clouds, heavy rain, and high winds hampered navigation. In order to pinpoint their position, the aircraft were forced to drop down to 200 feet where they encountered heavy flak from the German defenses. The parachute drop was disastrous, with the men being scattered over a wide area on landing and several injuries being sustained—Stirling himself was knocked unconscious. In addition, all of the teams had been dropped way off target.

The operation was abandoned as a result. Individual soldiers made their own way to the rendezvous with the Long Range Desert Group (LRDG). The problems encountered on this first operation prompted a radical rethinking about transporting troops to the target on future SAS missions. The main problem had been the weather. As fate would have it, the SAS had chosen to parachute in on a night when the weather was described as the worst in the region for thirty years. Notwithstanding, Stirling resolved never to use parachute drops again, preferring instead to use vehicles, starting with those of the LRDG.

The provisional war establishment of the LRDG was authorized in July 1940, originally for 11 officers and 76 men. This number was increased to 21 officers and 271 men in November 1940. By March 1942, the LRDG numbered 25 officers and 324 men. Operating in open-topped Chevrolet trucks, the LRDG carried out reconnaissance, intelligence gathering, and courier duties. These tasks involved long drives across the featureless desert terrain, followed by observation of enemy posts and convoys.

Long Range Desert Group penetrated hundreds of miles in trucks and jeeps carrying repair facilities with them. They transported and collected the SAS after the first early raids.

It was immediately after the first parachute disaster that the idea of using trucks came to David Lloyd-Owen, an officer with the LRDG, who at the time was discussing the failure of parachute entry with Stirling. Lloyd-Owen argued that the LRDG was already familiar with desert travel and navigation, and urged Stirling to consider the possibility. Stirling discussed this with the LRDG commander, Colonel Guy Prendagast, who agreed. So from late 1941 until early 1942, the SAS used LRDG trucks as a method of entry. Although Stirling's men caused problems for the LRDG because they provoked greater enemy reaction, the cooperation between the SAS and LRDG remained excellent.

After the failure of the first mission, Stirling had to face the music from MEHQ. Luckily for him there were a lot of changes being made at the time. The results of one small disaster by the SAS paled into insignificance beside the fact that the war against Rommel was not going well. During his visit to Cairo, Stirling kept away from MEHQ and did not surrender a report on the failure; as far as they were concerned the SAS was still in action—somewhere.

During this time, the SAS moved into new accommodations at Jalo Oasis, which had just been captured from the Italians. The oasis, which had originally been a French Foreign Legion outpost, became home to the LRDG and the SAS during the North African campaign of the Second World War. It changed hands several times during Rommel's push, but was finally taken from the Italians on November 25, 1941. With the aid of LRDG transportation, the SAS launched many of their early raids against the German airfields from Jalo.

Despite the shaky start, Stirling continued with his planning, and a month later several parties raided the airfields at Sirte, Nofilia Agheila, Agedabia, Tamit, and Arco dei Fileni (known as Marble Arch to the British), destroying some ninety-seven enemy aircraft. This was the success that the SAS needed as it would be hard for anyone foolhardy enough at MEHQ to criticize such a success.

The celebrations were short lived as Stirling ordered Fraser and Lewes to carry out renewed attacks on the airfields at Nofillia and Marble Arch, while he and Mayne attacked Sirte and Tamit. Stiling and Mayne left with their parties on Christmas Eve with Fraser and Lewes leaving on Christmas day.

On December 28, Lewes's party, consisting of Jim Almonds, Jimmy Storie, Bob Lilley, and Robert White, was dropped off by the LRDG some 18 miles from Nofillia. The raid had gone according to plan but proved to be ineffectual as the airfield was abandoned. Lewes and his team destroyed the only two aircraft they could find before making a rapid retreat. Some success was achieved when the frustrated raiding party came across a large number of Italian vehicles near a roadhouse at Mersa Brega, which they destroyed. The group then headed for the RV (rendezvous) with the LRDG, which they made it to successfully. However, as they traveled across the desert towards Marble Arch, they were attacked by a Messerschmitt 110 and Jock Lewes received a burst of gunfire in the back. He died of his wounds and was buried where he fell. The rest of the party was then harassed by Stuka dive bombers resulting in all four LRDG trucks being destroyed. However, one of the LRDG members managed to cannibalize the wrecks and make one working vehicle; by sundown of December 31, they had started the long journey back to Jalo Oasis.

Stirling's concept of small raiding parties proved to be a great success when on the night of January 23, 1942, the SAS struck at a German oil supply depot in Buerat, west of Sirte.

Jalo Oasis, there is an SAS jeep in the center of the image.

They destroyed eighteen enormous fuel transporters, each loaded with 4,000 gallons of fuel. At the same time they attacked the dock facilities and accompanying warehouses filled with everything from foodstuffs to heavy machinery . . . all this just prior to a major offensive by Rommel.

Around this time both the SAS and LRDG moved to Siwa Oasis, which was situated approximately 600km west of Cairo; Siwa was the main base of the Long Range Desert Group (LRDG) in North Africa. In June 1942, the SAS mounted a series of raids on enemy airfields from Siwa to help a convoy of seventeen ships attempting to reach Malta. Despite the massive diversions, only two ships got through. Around the end of June, after most of the raiding parties had arrived back at Siwa Oasis, Stirling was once again back in Cairo. This time it was to attend the wedding of Captain William "Bill" Cumper and his wife. Cumper was a Royal Engineers Officer and an expert in explosives. After the death of Jock Lewes he took over the explosives training for the SAS; he also trained the newly arrived Free-French in demolitions. Once the ceremony was over he was soon on his rounds of chasing and purloining everything he needed, and as luck would have it, he succeeded in getting his hands on a fleet of three-ton trucks and fifteen American jeeps.

Author's Note: The American Willy's jeep, much modified, was used by the SAS during the Second World War. It was light, agile, robust, and ideally suited to desert conditions because of its four-wheel drive. The jeep had a top speed of 60 mph and a range of 300 miles, although this could be greatly extended by carrying additional fuel. During the North Africa campaign the SAS jeeps were normally overloaded with extra water, fuel, and ammunition. The SAS also added some heavy firepower in the form of 0.5in Browning heavy machine guns and Vickers "K," the latter of which had been found disused in an aircraft hanger. So equipped, the jeeps were not simply used as a means of transport but also as a weapon. On many occasions they were driven, guns blazing, directly into the enemy airfields where they would wreck havoc on the rows of enemy aircraft. Between 1943–1945, the SAS used their jeeps in Europe, many of which were parachute-dropped to the teams on the ground. This gave the SAS the speed and flexibility to operate against the Germans.

The jeep raiding tactic has been used by the SAS ever since and remains one of its major modes of insertion, albeit in a more modified vehicle.

One such raid, on the night of July 7, 1942, led by Major David Stirling and Captain Paddy Mayne, proved extremely successful. After an overland approach in vehicles, the initial attack on the airfield was carried out on foot by a group of four men under Captain Mayne while Major Stirling established a roadblock nearby to attack enemy vehicles. Lewes's bombs were placed on a total of forty aircraft, but only twenty exploded because of faulty primers. As a result, Stirling decided to re-enter the airfield this time using the jeeps as weapons platforms. He drove down the line of aircraft with Jonny Cooper firing the single Vickers K directly into the planes. A further fourteen aircraft were destroyed by Stirling's "Blitz Buggy" in three jeeps. Reconnaissance later confirmed that thirty-seven planes had been destroyed at Bagoush, while Stirling and his men withdrew without loss.

At the same time, Captain Jellico led a raid on El Daba using jeeps in a hit-and-run attack. They failed to penetrate the perimeter and had to console themselves with blasting through the wire. The second raid four days later proved equally difficult as the Germans had upped their defenses, causing the attack to be called off.

SAS Jeep. The mobile firepower platform that served the SAS so well during WWII still remains a major form of infiltration and combat vehicle. The original show here demonstrates both the single and twin Vickers K machine gun.

Fraser and Jordan's parties drove close to the two Axis airfields known as Fuka Main and its satellite before approaching on foot. Fraser had returned to the RV around 2:30 p.m., having found Fuka Main too well guarded. Jordan had more success and had destroyed eight aircraft at the satellite field. Fuka Main was raided again on the night of July 12; this time the SAS managed to destroy 22 aircraft.

On July 26, 1942, the SAS raided an Axis airfield in the Fuka region of North Africa. Under cover of darkness David Stirling led fourteen jeeps, armed with Vickers "K" machine guns, split into two columns of seven commanded by George Jellicoe and "Paddy" Mayne, onto the airfield in formation. The SAS had riddled the parked aircraft with gunfire and in just a few short minutes they had destroyed forty aircraft before disappearing into the darkness. SAS casualties were small with the loss of only one man killed and two jeeps destroyed.

Typical German airfield in North Africa which was raided by the SAS.

By providence, a paper written by General Alexander's Chief of Staff, recommended that, due to the effect of Special Forces working behind enemy lines, L Detachment SAS Brigade should receive Regimental status. Churchill is said to have been enthusiastic and added his backing for the Regiment to become a recognized part of the British Army. On September 28, 1942, Operational Instruction No 14,521, 1 SAS was officially listed and had amalgamated into its ranks both the Special Boat Section and the 114 men of the Greek Sacred Squadron. Stirling himself was promoted to the rank of Lieutenant Colonel but his independence was diminished as 1 SAS now came under the control of the Director of Military Operations.

Author's Note: It was not just the British who realized how effective the SAS had become. On October 18, 1942, Hitler signed a Commando Order in response to the many raids and sabotage missions carried out by British Commandos in North Africa. The German dictator issued a directive in

October 1942 stating: "I therefore order that from now on all Allied Commandos encountered by German forces in Europe and Africa should be killed immediately, even if in uniform or if they attempt to surrender. . . those in so-called Commando operations. . . are to be exterminated to the last man in battle or flight. Even should they, on their being discovered, make as if to surrender, all quarter is to be denied on principle." The order made it clear that failure to carry out these orders by any commander or officer would be considered to be an act of negligence punishable under German military law. This obviously gave local commanders license to execute any captured Commandos, including SAS, and during the French campaign in 1944 the Gestapo, in response to Hitler's order, executed several SAS soldiers. At the Nuremberg Trials, the Commando Order was found to be a direct breach of the laws of war, and German officers who carried out illegal executions under the Commando Order were found guilty of a war crime.

Not long after, Stirling himself was captured. But instead of being shot, he was taken to a Calvary Barracks in Rome and later imprisoned at the notorious Colditz Castle until the end of the war—a lucky man if you consider the fifty officers who were murdered for their repeated escape attempts (depicted in the film *The Great Escape*) and the high number of Stirling's own men who disappeared at the hands of the Germans under Hitler's Commando Order.

The capture of David Stirling caused great confusion within the SAS, and his long-standing enemies at MEHQ saw this as an opportunity to bring the regiment into line. Command of the regiment immediately went to the most senior officer, Major Vivian Street, OC B Squadron, but he was inexperienced in SAS operations. Shortly after, Lieutenant Colonel H. J. Cator, once commanding officer of 51 Commando, was appointed, but on March 19, 1943, the 1st SAS Regiment's 47 officers and 532 other ranks were split into three groups.

Author's Note: Today, as part of SAS Selection, successful candidates are tutored in the art of "Resistance to Interrogation." Understandably, one of the most frightening experiences a soldier can face is being captured and interrogated by the enemy, and members of the SAS, by the very nature of their methods of operation, are particularly vulnerable to capture.

As a result, all SAS soldiers are required to undergo JSIU (Joint Services Interrogation Unit) training.

Interrogation scenarios are made as realistic as possible, although torture during training is not allowed. However, controlled applications of fear-inducing methods are used, such as white sound and the wearing of pillowcases soaked with water to give the impression of slow drowning. The rule is simple: give your name, rank, and number only. The period of interrogation and isolation lasts for twenty-four hours, and during this time the prisoner is constantly hooded and kept in a damp, cold room.

The Geneva Convention states that every member of the armed forces, should he be captured, will only divulge his personal number, rank, full name, and date of birth. The enemy, on the other hand, is unlikely to stand by the Geneva Convention, and will use any one of hundreds of known interrogation techniques to "persuade" a prisoner to volunteer information of intelligence value. In fact, they are more likely to use the convention in an attempt to trick prisoners into collaboration, which can only be successfully avoided if the prisoner is fully aware of his rights under the Convention. All SAS prisoners of war recognize just how imperative preserving military security is, as careless information could easily result in the capture or, at worst, the death of others. Three members of the SAS were brutally tortured by the Iraqis when captured during the Gulf War, but as we know, gave no information whatsoever, thus proving the enormous value of interrogation training.

AEGEAN SEA

After their success in North Africa, the SAS moved on to Sicily and the Aegean. The Aegean features the Dodecanese, a large group of islands which were the scene of operations by the Special Boat Squadron (SBS), commanded by Major Earl Jellicoe. These operations were carried out against German forces in occupation on a number of islands including Crete, Karpathos, Leros, Kos, Simi, Tilos, Astipalaia, and a number of other smaller ones, which formed part of a defensive perimeter against Allied operations in the Mediterranean.

In January 1944, the SBS established forty heavily camouflaged anchorages which served as bases for operations among the Dodecanese Islands. A schooner, the *Tewfik*, acted as a forward headquarters for Major Jellicoe while a fleet of armed caciques, motor launches, and other small craft, under the command of Lieutenant Commander Adrian Seligmann RN, was employed to ferry the SBS squadrons around the islands. Overall control of raiding operations in the Aegean was the responsibility of Headquarters Raiding Forces, based in Cairo, which was formed in late October 1943, under the command of Colonel (later Brigadier) Douglas Turnbull.

Author's Note: One of the most famous members of the SAS, and one who remains a personal hero of mine, was a young Dane called Anders Lassen. Lassen joined the British Army at the beginning of World War II and participated in commando training before joining the specialist Small Scale Raiding Force. In May 1943, he was in the Middle East as part of D Squadron, 1 SAS, which became part of the Special Boat Squadron. His fighting qualities soon came to the fore, and Lassen was soon recognized as being an exceptional soldier, ideally suited to elite forces operations, where quick decisions and raw courage are often called for, combined with ice-cool nerves. He took part in a number of SBS missions, such as the raids against Crete and Sirni in June 1943, while in 1944 he fought in Italy, the Adriatic, and Greece. It was in 1945, though, that he entered SAS legend. Fighting around Lake Comacchio, he was killed while attempting to destroy a number of German pillboxes. Before he died he had captured or killed a large number of Germans, and his presence had even affected enemy strategy in the area. For this and his actions he was awarded Britain's highest honor: a posthumous Victoria Cross.

EUROPE

As "D-Day" approached, the SAS prepared to enter the war in Europe. Their main task was to organize and help the French resistance and to frustrate the Germans by snapping at their heels behind the lines. This was highly successful but was also done at a high cost, as reprisals were swift and savage. The first of many such reprisals happened during Operation "Loyton." Lieutenant-Colonel Brian Franks was to lead some ninety-one men of 2 SAS; their mission was to strike at enemy installations and cooperate with the local maquis (which was problematic at the best of times).

They would be situated in the area of Vosges in eastern France. The advance party, together with a phantom patrol and a Jedburgh team, were dropped on the evening of August 12, 1944, with subsequent drops taking place in the days that followed. Unfortunately, the Germans had placed large numbers of troops on the crests of the Vosges and the east bank of the River Moselle. The SAS party thus parachuted into an ambush. To make matters worse, the Gestapo were also in the area—Nancy and Strasbourg—and both locations had anti-partisan units. The SAS men had to contend with large numbers of enemy troops and traitors within the maquis. There were also a number of reprisals against local villages in response to SAS successes.

Author's Note: The word Jedburgh is taken from a small Scottish village. The teams were raised to work as part of the SOE and American OSS. A team comprised three men: a leader, an executive officer, and a non-commissioned radio operator. The teams could be made up from a variety of nationalities: British, American, and French—there was not a set rule. The days of sending messages by pigeon were replaced with the "Jed set" or the type B mark 2 as it was officially designated; this was the main communication device used to contact SFHQ in London.

A total of ninety-one Jedburgh teams were dropped into France; the first parachuted into central France near Châteauroux the night before the Allied landings in Normandy and the invasion of Europe. Most insertions were at night and they would be met by a reception committee from a local Resistance cell. Their mission was to forge a link and command structure among the various Resistance groups and supply them with arms and ammunition. Once this was done, they would help direct the Resistance movement to aid the war effort of the Allies.

As with the SAS, those members of the Jedburgh teams who were captured were subject to Hitler's Command Order and therefore subject to torture and execution. Fortunately, of the French Jedburghs, only British Captain Victor A. Gough met that fate. He was shot while a prisoner on November 25, 1944.

But it was not just the soldiers that suffered. Any civilian found harboring or helping the Allies also paid the ultimate price. Moussey is a small valley town near the Vosges

Mountains of eastern France, 40 miles southeast of Nancy and ten miles north of St. Die. The village itself is spread out for about a mile, with odd houses dotting the roadside. Towards the center is a church with a military graveyard where men from 2 SAS are laid to rest.

The unit parachuted into France in September 1944, landing north of Moussey near Baccarat. It was to be one of the last drops the regiment did during the Second World War. Soon after the drop, the group moved to the woods and forests close to Moussey. The local population befriended the unit and as time went on, provided them with the necessities of life and with whatever help they could. The American Army, under General Patton, was held up owing to lack of supplies. It was a delay the Germans took advantage of, moving reinforcements along the River Meurthe, a few miles to the west of Moussey. Numerous raids were carried out on the Germans, which in the end brought retaliation. The male population of Moussey, for example, was rounded up and packed off to concentration camps. Of the 210 men between the ages of sixteen and sixty who were taken, only seventy returned. Despite this, the SAS headquarters, hiding in the nearby woods, was never discovered. Franks brought the

A postcard showing the village of Moussey before the Germans took reprisals for locals protecting the SAS who had been hiding in the nearby woods.

169 — MOUSSEY
Vallée du Harcholet

operation to an end on October 9, by which time two SAS soldiers had been killed and thirty-one captured (all of whom were shot by the Gestapo). The lack of supplies and erratic resupply drops had brought the mission to an end.

Author's Note: The SAS suffered thirty-two casualties during the whole operation, with ten falling close to Moussey. In the late summer of 1945 the villagers of Moussey created a small cemetery in their church grounds—the whole village attended. Some years later, and at the behest of the village, the War Grave Commission designated the Moussey graveyard as an official war cemetery, thus binding Moussey into SAS history forever. A representative of the SAS still visits on an annual basis.

As the war began to wind down, the SAS Brigade began to disintegrate. The British SAS elements found themselves doing the oddest of jobs such as Operation "Apostle." HQ SAS Brigade, consisting of 1 and 2 SAS, all under the command of Brigadier Mike Calvert, had to disarm the three hundred thousand German soldiers remaining in Norway at the end of World War II. The two regiments were shipped from Ostend to Tilbury, and in England they were issued with new jeeps, new clothing, and equipment.

An advance party made up of representatives from both regiments and a detachment from Brigade Staff arrived at Stavanger on May 12, 1945, and by the end of the month, a total of 760 troops—17 motorcycles, 68 trailers, and 150 jeeps—were in the country. The SAS Brigade was based at Bergen to administer the operation. The four months that the SAS soldiers spent in Norway were hardly taxing, and most of the men regarded the time as a sort of vacation. The Germans gave no trouble (although there were clashes with Quislings—Norwegian collaborators); the locals were largely friendly and weather was warm. The SAS returned to England at the end of August. An amusing footnote to the operation is the so-called Battle of Bergen. The young ladies of the town were quite fond of the daring British soldiers in their midst, a fact resented by many young Norwegian males in the town as well as members of the local police force. This culminated in a large-scale brawl in the town's center, which the SAS won hands down. The "victory" resulted in British diplomats in Norway urgently requesting the War Office to evacuate the SAS to England.

One final task assigned to the SAS fell to Major Eric "Bill" Barkworth, an intelligence officer of 2nd SAS Regiment. A War Crimes Investigation Team was formed in May 1945 by Lieutenant Colonel Brian Franks, the Commanding Officer of 2nd SAS, to investigate the fate and whereabouts of thirty-two men missing from the unit. The men had been deployed in October 1944 on Operation Loyton in Alsace, and it had been reported that they had been captured and executed by the Germans.

The small six-man team was under the command of Major Barkworth, with WO2 (SSM) Fred "Dusty" Rhodes as his second-in-command. At the end of May, Barkworth and his men set off from England and traveled to Germany. The team was in direct contact with the SAS headquarters at Wivenhoe, in Essex, via a radio link maintained by F Squadron Phantom. On June 10, the team reached Gaggenau where a number of bodies had already been discovered near Rotenfels Concentration Camp and exhumed by French occupation forces. Some of these were identified as 2 SAS personnel. Reinforced by a few more members of the regiment sent from England, the team widened its search.

In the late summer of 1945, as a result of the disbandment of the SAS, the team was sent directly to the War Office. Thereafter its radio base station was located in the offices of the Department of the Judge Advocate-General where Captain Yuri Galitzine looked after the team's affairs and provided any necessary support. In November, the team turned its attention to the Moussey area of Alsace and after lengthy and exhaustive searches, uncovered a mass grave containing eight bodies which were subsequently identified as being those of 2 SAS men. Further investigations in the area uncovered the murders of other members of 2 SAS.

After six months, the team had amassed a considerable amount of information which enabled them to account for all the missing men. The team was also successful in tracking down their executioners who were arrested and having been brought to justice, were hanged at Hameln prison. The team also assisted in the investigation of other war crimes involving the executions of Allied aircrew and four women agents of the American Special Operations Executive who had been murdered in the concentration camp at Natzweiler.

Natzweiler concentration camp gates—one of the last places visited by the SAS war crimes team.

Finally, while much of the British Army was disbanded, the SAS would remain and survive as a territorial unit. This in turn would form a small unit which was designed for the war in Korea; instead they found themselves in Malaya as a unit known as the Malayan scouts from which modern-day 22 SAS evolved.

SUMMARY TACTICS & OPERATIONAL SKILLS

The skills that arose from the foundation of the SAS laid the foundation for the future. The only difference being in 1942 was a matter of "make do with what you have available and adapt," whereas today they are equipped with the most sophisticated equipment and weaponry available. Most importantly, Stirling learned to select the best men for the job. Some of the tasks carried out in 1942 were similar to those still being done by modern SAS today. The jeep raids behind German lines and the fighting columns operating inside Iraq are just two examples.

Likewise, due to the vast distances they would have to travel in order to complete their missions behind the German lines, Stirling needed a method of entry (MOE). At the time, military parachuting was new and as such untested. Jumping off the back of a truck to simulate a parachute landing was entirely different than actually landing with a parachute. However, Stirling was quick to recognize the dangers of this method of entry and adopted a safer approach by using the LRDG. This was one of the first tactical lessons the SAS had learned.

A quick and simple method of destroying an aircraft was required, and again the SAS produced possibly its first original piece of specialty equipment—the "Lewes bomb." They also found the best place to attach it in order to complexly destroy the aircraft and leave behind no usable parts. Eventually the Lewes bomb developed into the SAS "standard charge."

While the partnership with the LRDG worked, the SAS wanted their own transport; finally this came in the much needed shape of a heavily modified American Willy's jeep. Fitted with long-range fuel tanks and equipped with aircraft Vickers K machine guns, this became an SAS fighting platform for the whole of the Second World War. Henceforth, Mobility Troop became a reality which exists to this day.

Chapter 2

SAS SELECTION

It should be made clear that the SAS are not machines; neither are they a clinical force cloned for warfare. They are what they have always been: a collection of dedicated soldiers. These are men who love military life and wish to reach the peak of their chosen profession. Above all they are men who have found the true meaning of self-discipline. In truth they seek adventure in the SAS and thrive on the adrenaline rush, experiencing the same ecstasy as those who participate in high-risk sports. The standards for entering the SAS have been honed ever since the regiment's conception, and standardized since the mid-1950s. To reach this standard, every soldier must pass SAS Selection, possibly the stiffest military enlistment course in the world.

SELECTION

The present SAS regular regiment is based near a small village called Credenhill, near Hereford. The modern day 22 SAS came into being because of the communist troubles in Malaya between 1950–1959. In the early 1950s a British officer, Mike Calvert, who was serving in Hong Kong at the time, was instructed to apprise the communist influence in Malaya. Calvert, a tough soldier who had commanded an SAS brigade during the last war, was the innovator of the Malayan Scouts (SAS). His basic doctrine influenced many of the SAS practices that still exist to this day, the most famous being a four-man patrol and the development of individual skills. When the Malayan Scouts became formally known as 22 SAS, Calvert instructed John Woodhouse to return to England and set up a structured selection course, the basis of which remains to this day.

All SAS soldiers have passed Selection which makes them members of a unique organization; it also makes a new member acceptable to the older members of the Regiment.

The SAS Selection course takes place in the Brecon Beacons. The whole course has one aim: to weed out those who are unsuitable and

Would-be candidates climbing up Brecon Beacons in South Wales, a small mountain range that has claimed the lives of many attempting SAS Selection. In 2014, three candidates died during same course. The orange patch atop the bergen is for easy identification.

to push to the limit those who pass. The overall course, which was devised in 1953 by Major John Woodhouse, has changed very little over the years (although more emphasis was placed on safety after a series of deaths in the late seventies and early eighties). The course is long and tough. Those who pass the "build-up to test week" find at the end they are faced with the "Endurance March." Little can prepare one for this challenge; to succeed inside the allocated time is a fitting achievement in itself.

SAS Selection is hard; there is no other way to put it. The basis of the selection system is to ensure that valuable training time is only spent on the very best candidates. Ideally, candidates for the SAS need to have had at least three years service with a parent unit of the armed forces. This ensures that the basic training requirements and disciplines are already in place. In reality, most of the candidates come from the Airborne Regiments and elite Infantry units—but no one is barred from applying. Generally, those who do volunteer are men who love the military world—men with the physical stamina of an Olympic athlete—men who have found the true meaning of self-discipline.

Selection is perhaps the wrong word to use. The verb "select" means to pick out the best or most suitable, while the adjective ("selected") means chosen for excellence. The problem is that nobody picks or chooses the candidate; he must earn his place. It is more a case of the individuals selecting themselves. It is the single factor which makes the SAS so unique: a whole bunch of individuals, but with the capacity to act as one. All candidates who pass selection must give up all previous rank held in their original regiments and revert to being troopers (known as Blades). Each individual must then work his way back up the promotion line. In fairness, they keep their original pay, and after a specified period they also receive additional SAS pay—for a senior rank (Staff Sergeant) this can be as high as £80,000 a year when benefits are included.

The physical side of selection takes place in and around the Brecon Beacons in South Wales. Although not a high range of mountains, they are treacherous, exposed, and battered by constant weather changes; death by hypothermia or heatstroke is seldom far away. In the spring of 2014, three candidates died attempting SAS Selection. This is not new, as numerous SAS candidates have suffered a slow death, while lost and disorientated or simply by pushing their bodies beyond safe limits. It is therefore essential that a diligent, self-imposed training schedule be undertaken by the candidate prior to arriving at Hereford.

SFBC

The SAS run the Special Forces Briefing Course (SFBC) to ensure that prospective candidates are fully aware and prepared before they attempt Special Forces selection. It is also an opportunity for the regiment to look at prospective candidates, making sure they like what they see.

Candidates will be given a series of briefings and presentations about the role of British Special Forces in general and specifically that of the SAS. This normally starts with a talk from Training Squadron OC, prior to being briefed on what selection is all about and how best to prepare themselves. An example of an SFBC weekend is outlined below.

Friday Evening

The candidates are given a brief as to what is expected of them. This is followed by the standard British Army totes and a map reading test. There is also a map memory test as within the SAS marking a map in any shape or form is strictly forbidden outside of the Operations Room. Due to the severity of the process and the number of deaths during selection, there is also a brief on emergency First Aid and how to deal with hypothermia and heatstroke (depending on whether the selection course takes place in winter or summer). Candidates are also tested on their military knowledge as well as being given an IQ test.

Saturday

Before breakfast, candidates will do an APFA and a Bleep Test, after which they will be taken to the swimming pool where they will be required to jump from the high diving board. This is to assess their initial aptitude for parachute training. During the visit to the pool they will also be required to swim 100 meters in three minutes, after which they must tread water for ten minutes. All this is done in combat clothing and trainers. For the rest of the morning, candidates will receive briefings on the regiment, which include not just the combat side, but also subjects such as welfare and daily life at Hereford.

In the afternoon they will be driven to a training area to the south of Hereford, where they will carry out a series of fitness tests; these include the British Army standard fitness test. This consists of a static lift, water container carry, and a run.

Static Lift

The Static Lift is described as an exercise to simulate lifting heavy items such as kit and ammo onto the back of an Army truck. The reality is that you will be expected to lift power bags which will vary in weight, progressively becoming heavier, safely to a height of 1.45 meters.

The weight of each power bag ranges from fifteen to fifty kilograms (33 to 110 pounds). They must lift each bag in order of weight until completion or failure, and their score will be based on the total amount of kilograms lifted.

Jerry Can Carry

This test is to determine the strength of the upper arms and shoulders. It is also a test of grip. Candidates are required to carry two jerry cans (water containers), each weighing 20kg (forty-four pounds) along a total distance of 150 meters. With arms at their side and carrying one jerry can in each hand, candidates are expected to complete this course in under two minutes. They are required to keep a pace of no less than 5.4 kmh (3.36 mph) and will be scored on the distance in meters that they can carry the 20kg weights while maintaining the minimum pace.

The Run

The famous run is a 2.4km (1.5 mile) track in which they must complete the full distance within the given time. The time given to complete the run will vary depending on the position within the Army that candidates have applied for.

Before starting the timed run, candidates will warm up as a squad with the other people who are also in the selection process. This consists of a slow jog/walk over a distance of 800 meters. They will then immediately begin the test. The required times vary regiments within the Army, but for the SAS, they adopt that of the Parachute regiment, with a run time of 09:40 min.

All these tests are done back-to-back. Saturday ends with a briefing on the activities of Special Forces, detailing some of their specialist roles.

SUNDAY

The DS will run the candidates for about one and half hours, during which time they will be required to carry other candidates. These distractions will include both the fireman's lift and baby carry (cradled in your arms), feats the candidate must do going both up- and downhill. Candidates then return to the barracks and watch the SAS Regiment video, which gives a rather glamorous laid-back image of the SAS. Before dispersal around lunchtime, candidates will undergo a final interview.

To help reduce the numbers attending SAS selection, only the very best are invited to return; those thought to be lacking in certain aptitudes are required to brush up on their weak areas and return again when ready.

Test week comes at the end of the first phase of SAS Selection. The week consists of a series of tests designed to push the individual to the very limits of endurance. Hence the name of the final test: "Endurance March." Little prepares the candidates for this. With a rifle and bergen (rucksack) weighing 25kg (55 pounds), they are expected to walk 40 miles in just twenty hours. No problem, one might say, until they realize the route runs up and down the Beacon's mountains. The Endurance March starts early in the morning, and the candidate needs to make good time by maintaining a steady pace—a walk uphill, a run downhill. Energy expenditure is high and food stops are required every three hours, but these rest periods, which can last twenty minutes, burn into the overall time. Most make a brew of tea and eat a small amount of high calorie, light food. Eating too much food will cause the candidate to vomit. Time during the rest break is spent checking the map and memorizing the route. During summer selection, the candidate must watch for signs of salt deficiency through sweating, while in winter they must guard against hypothermia. Simply being too hot or too cold will severely affect the individual's performance and overall route timings. Irrespective of what the weather looks like at the outset, the candidate should always carry a complement of clothing that will cater for all climatic conditions. The weather over the Brecons can change rapidly, from sunlight to thick fog. Accurate compass work is needed in order to prevent the candidate from staggering around aimlessly when visibility is at arm's length. In addition to map and compass navigation, the candidate also requires a working knowledge of the Global Positioning System (GPS). For the lucky candidates who find their way to the final RV within the allocated time, they can rest assured that, physically, the worst is over. Yet there is no time for relaxation as they still have a long way to go before entering into the ranks of the SAS.

Continuation Training where basic and new skills are taught. Learning to operate in a small patrol is one of the first requirements.

Continuation Training follows, and all the necessary basic skills required of the SAS soldier are taught and practiced. For many of the Para's and Infantry guys, it's back to basics: weapon training, patrolling skills, SOPs (Standard Operational Procedures), escape and evasion exercises, parachuting, and finally five weeks jungle training. For those who cannot swim or drive, there is a crash course in both before the candidate is allowed to enter an SAS Squadron. Finally, before training is done, he will receive his beige beret with its famous winged dagger—as any SAS man will tell you, it is a special moment and a fabulous feeling!

Author's Note: As so many retired members of the SAS have become settled and married to local women, it is understandable that some retain their affiliation by becoming members of L Detachment (previously known as R Squadron). L Detachment is a territorial (reserves) unit directly attached to 22 SAS in Credenhill, made up of retired Squadron members and civilians. All civilians in L Detachment have to go through the same selection process as the regular squadrons. Its role is to augment the four regular squadrons, providing additional manpower as and when needed. Members of L were first sent into action in 1991 during Operation Desert Storm, when they bumped up the numbers in A and D Squadron Land Rover columns. These civilian soldiers, operating behind Iraqi lines, did extremely well and showed great courage.

Survival and Resistance to Interrogation Training is a vital part of SAS Selection. These men are dressed in the traditional old army dress coat tied up with string. They are hunted for a week, living off the land before being "captured" and interrogated.

SURVIVAL TRAINING

The SAS take Combat and Survival training very seriously, and they start the process during selection. The reason is simple: any SAS operation involves action behind enemy lines or in remote unfamiliar terrain where the risk of capture and the need to survive are ever present. The subjects involved are fascinating and are all taught by both military and civilian experts. The aspects cover escape, evasion, survival, and resistance to interrogation.

The course combines classroom lessons, practical outdoor instruction, and a final exercise. Even if candidates avoid capture they will still have to undergo the compulsory twenty-four-hour interrogation phase. This will involve a very realistic examination of the candidate's ability to resist questions put to him under duress. Joint Services Interrogation Unit carries out all SAS interrogation training.

INTERROGATION TRAINING

Because the SAS generally work in a hostile environment, normally way behind enemy lines, there is a strong danger of getting captured. While most SAS soldiers will do everything in their power to avoid capture, sometimes it is inevitable. To help the soldier resist interrogation, they are subjected to a twenty-four-hour period of training. This training is carried out by Joint Services Interrogation Unit (JSIU), which is part of the Intelligence Corp. The JSIU not only puts SAS soldiers through a rigorous training regime, but are often on hand in the local theater of operation to interrogate prisoners captured by the SAS.

For those soldiers just finishing "Selection and Continuation Training," resistance to interrogation usually comes at the end of a long Escape and Evasion exercise. This means they will be cold, hungry, and somewhat softened up—ready for the process. This involves being tied up and hooded. Next they are forced to stand against a wall for hours at a time before eventually being taken before an interrogator. Interrogators can be warm and friendly or absolutely terrifying. They will do anything to get them to say more than what the Geneva Convention stipulates (i.e., number, rank, date of birth, and name—known as the big four).

Around the twelve-hour mark, they try a variety of ploys, from hosing the men down with water, to playing "white sound" and good cop bad cop routines. As long as the candidates stick to just saying the following: "I'm sorry sir, I cannot answer that question" and giving them the big four, they will be OK.

While interrogation training does help, being in the hands of real enemy interrogators is a totally different ball-game. First off they are likely to get beaten badly by the soldiers that capture them—and most probably shot. If they are taken prisoner and the enemy half guesses that they are Special Forces, then they're in for a rough time. The aim is to resist for the first twenty-four hours, giving other members of their patrol time to reach the RV and escape.

If they are undiscovered and running uncaptured, the SAS are issued with several items to aid them reaching safety. These include an Emergency Beacon, which allows any user to contact most friendly aircraft and thus raise the alarm and request pickup. For more remote and inhospitable environments, each SAS soldier carries an Escape and Evasion kit.

ESCAPE KIT

The essentials of an SAS escape—for most SAS soldiers—are housed in a tobacco tin, the contents of which will depend on the theater of operation. Items can include a button compass that can be swallowed, a wire saw that will cut through the hardest metal, a condom for collecting water, and a tampax for use in fire lighting. In addition to the basic survival items, specialist equipment is also issued such as a cloth escape map, printed on silk, or lock-picking tools.

A typical SAS Survival kit will therefore contain the following:

Button compass, wire saw, wind and waterproof matches, flint and steel, safety pins, candle, signal mirror, fishing kit, puritabs, razor blade, condoms, snare wire, tampax, poly-ethylene survival bag.

Blood Chit

The blood chit is basically a piece of paper that an escaping soldier can show to any civilian that may help him. Each blood chit has a unique number at the top; any civilian who has aided a soldier may approach any British Embassy or Consulate and claim his reward. This blood chit was issued for the Gulf War and is in English, Arabic, and Farsi.

Blood Money

Gold half-sovereigns are issued to SAS soldiers who operate behind enemy lines. The purpose of such money is to back up the blood chit, since not everyone can read. Additionally, gold is recognized the world over for its value. All SAS soldiers in the Gulf War were issued with twenty gold half-sovereigns, and many were used in order to effect a safe return to their own lines after getting separated from their unit. In some parts of the world American dollars are used in the same way.

JUNGLE TRAINING

Jungle training is an important requirement for the SAS, as they are often committed to operations in tropical regions. Surviving in the jungle is not easy; patrolling and operating in secret require special skills. SAS jungle skills are designed so that a small team is able to operate, fight, and survive in a hostile environment. Moving from place to place in the jungle is slow and laborious, and there are hidden dangers all around, not just from the enemy, but from a host of different insects, animals, and the jungle itself.

In SAS selection, the jungle phase takes place twice a year, at the Jungle Training School at Tutong, Brunei. The school is run by experts from several nations, including several instructors from SAS Hereford, and the six-week training periods are run in March and September. The SAS has a saying that "the jungle is a great equalizer of men,"

English

I am British and I do not speak your language. I will not harm you!
I bear no malice towards your people. My friend, please provide
me with food, water, shelter, clothing and necessary medical
attention. Also, please provide safe passage to the nearest friendly
forces of any country supporting the British and their allies. You
will be rewarded for assisting me when you present this number
and my name to British authorities.

Turkish

Ben İngilizim ve sizin dilinizi bilmiyorum. Size kötülük
etmeyeceğim! Sizin halkınızdan nefret etmiyorum.
Arkadaşım, yemek, su, sığınak, elbiseler ve gerekli tıbbi
yardım bana verir misiniz lütfen? Ayrıca, İngilizleri ve
müttefiklerini destekleyen ülkenin en yakın güçlerine
tehlikesiz geçiş sağlar mısınız lütfen? Siz, İngiliz
yetkililerise bu numarayı ve benim adımı sunarak, bana
yardım ettiğiniz için ödüllendireceksiniz.

Armenian

[Armenian text – illegible]

Aramaic

[Aramaic handwritten text – illegible]

Kurdish

من بریتانیایم و زمانت تارتان   من شانه ویٰ زبان یات به دم ـ
من د زمانی خه لکیم ـ دوزسم، نکا ده کم حزری و تار و به نا و حل
و یارمه تی یی په زیشکی به من بده ی ـ مه روه ما من نکا ده کم به من
یارمه تی بدهی بوٰ نه وه ی بچم به سه لامه تی به سه دباره دوٰزمنی
نزیکاره کانی کام و رۆکت که به بریتانیاپیٰنه کاری لی ده سنه ان بدا
ته که ر نام زنبره و باری من به د مه سه لانی دریتانیای بدهی نز
یاغاسلیک بوٰ نام یارمه تیه ده مستۆسی

Persian

من انگلیسی هستم و زبان شما را نمیفهمم من به شما صدمه ای نمیزنم و نسبت به
مردمان شما هیچ و هنز ندارم فقط از راه دوستی خواهش دارم و به من آب و غذا و
مسکن و لباس بدهید و اگر لازم باشد کمک پزشکی و داروئی برایم فراهم کنید.
همچنین اگر برایتان ممکن باشد مرا از باد طریق ایمن و بی خطر به نزدیکترین
نیروهای طرفدار بریتانیا و متفقین آن کشور هدایت بفرمائید. اگر به من کمک کنید
کنید این شماره من و اسم من را به مقامات انگلیسی بدهید تا به شما پاداش
شایسته ای بدهند

Arabic

انا بریطانی لا اتکلم ولا فهم لغتك لن اسبك بازی لا اضمر الحقد والشر
لشعبك یا صدیقی لرجوك ان ترفق لی الطعام وماء وملجأ وملابس والعنایة الطبیة
الضروریة کما ارجوك ان ترافقنی الی اقرب قوة للقوات الصدیقة التابعة لایة دولة
تساند البریطانیین (الانكلیز) و حلفائهم اذا ذكرت وساعدتنی سعال مکافأة عند ما
اقدم هذا الرقم واسمی للسلطات البریطانیة.

Serial No.

Blood Chit issued to the British RAF during the war in Iraq. The SAS are issued with
a similar document.

meaning each soldier must find his own way of coping with the jungle environment. Many find the closeness of the forest and dense foliage very claustrophobic, while others enjoy the challenge, seeing it as an adventurous playground.

Jungle Training circa 1966; nothing has changed. Note that each man has his belt kit on and is actually holding his weapon.

Everything in the jungle is growing or dying together. There are a million smells and thousands of sounds from birds and animals. Jungle training involves learning jungle navigation and patrolling techniques where candidates are required to find their way through thick, dense foliage on steep-sided hills and over the most rugged terrain. The threat of coming face-to-face with the enemy is ever present, so close quarter contact drills and techniques of preparing a quick ambush are an integral part of the training. Jungle training is a tough discipline, which relies on the use of hand signals, the minimum amount of noise possible, slow movement, and testing routine.

The SAS will often be requested to undertake missions in any one of the world's jungles, all of which differ greatly in the risks they pose. In the Far East, Malaysia and Brunei, the jungle is reasonably safe, while that of South America, in particular Belize, is particularly dangerous with an enormous number of hazards. Here even the trees contain an acid-type sap.

TROOP SKILLS

Each Squadron is divided into four Troops and a small Headquarters section. The troops are designed to operate in all terrain and environments, providing the different methods of insertion, mobility, mountain, air-insertion, and boat. Each patrol within a troop is made up of four men; however, in recent years this has been increased to six due to the amount of communications equipment and firepower carried by each patrol. Within each troop, every SAS soldier will learn an individual skill such as medics, languages, demolitions, or signals.

MOBILITY TROOP

Mobility troop operates using a wide variety of vehicles, of which the SAS pink panther (Pinkie) is best known. The current vehicle is the 110 Land Rover, which comes fitted with a variety of armaments including GPMGs, .50 cal, mark 19 40mm grenade launcher, and .30 cal ASPs. They are all

fitted for long range and heavy engagements. Additional weapons include 80mm mortars and Milan Missiles. Other vehicles used by Mobility Troop include KTM 350 and Honda 250 motorbikes. Courses for members of Mobility Troop cover several weeks with the REME, doing basic mechanical fault-finding and emergency maintenance. Training in cross-country conditions can vary from the UAE (United Arab Emirates) to the deserts of America.

Mobility Troop came into their own when the fighting columns were developed for the first Gulf War. The motorbike was used as an outrider scouting for signs of the enemy which the main column could engage.

MOUNTAIN TROOP

Mountain Troop is responsible for all aspects of mountaineering and skiing. Training for the troop takes several forms. At times the whole troop may embark on a rock-climbing course where new members, with no previous experience, will be taught the basics of rock climbing and rappelling techniques. Depending on length of service and aptitude, many individuals will attend courses in Europe, with a selected few attending the German Alpine Guides course. This course is held at the German Mountain Warfare School in Mittenwald and nor-

mally lasts a year, with the Germans allowing two SAS personnel per course. The Mountain Guides course is divided into summer and winter mountain skills, and those that qualify return as both expert climbers and skiers. Over the past few years the mountaineering and skiing skills within the SAS have become highly proficient, and they now run their own Alpine training course. Additionally, several SAS members have even climbed Mount Everest.

Mountain Troop trained to a very professional level; this picture is of Sergeant Andy Baxter, just one of several SAS soldiers who died while climbing Mount Everest.

Mountain Troop is also responsible for ski instruction. Most troop members will be required to instruct the other squadron members during the annual winter exercises in arctic Norway. Again, several advanced ski instructor courses are available to the Regiment, in both France and Germany.

AIR TROOP

Air Troop is the "Free-Fall" troop within each squadron, whose normal role is that of pathfinder or, when operation necessity demands, covert parachute entry. They specialize in both High Altitude Low Opening (HALO) and High Altitude High Opening (HAHO).

Depending on the tactical circumstances, aircraft deploying free-fall troops are able to fly at heights of up to 35,000 feet. At this altitude oxygen is required and is either administered from a central console when in flight, or by individual masks during the drop. Each parachutist usually carries his equipment in a Bergen, which is attached upside down to the back of the thighs until it is released once the canopy is deployed. However, most free-fallers will hold the bergen with their feet until they are 50 feet from the ground, despite it being attached to the parachute harness by a 15-foot line, as releasing the bergen incorrectly can result in serious injury.

Air Troop member doing a HALO jump. Today all SAS members are trained in all methods of parachuting.

Ram-Air parachutes are normally used by free-fall troops since the rectangular design of the box allows for improved maneuverability and control, as well as an adjustable rate of descent. The front of the parachute is open to the driving air while holes in the internal sections allow for cross venting of the air, causing the parachute to inflate like a wing. Direction and descent of the canopy is controlled by two lines attached to the outer, rear corners of the wing. Pulling both control lines down fully literally stalls the canopy to a complete stop on landing. The glide ratio on a ram-air parachute is about 4:1.

Air Troop also participates in various other methods of flying, one of which has included the use of a small motor powered trike attached to a ram-air parachute. The vehicle was relatively easy to fly with a range of up to 180 miles, despite being particularly frightening at takeoff and landing.

The first free-fall operation carried out by the SAS was in the north of Oman, when two Air Troops entered the Wadi Rhawda under cover of darkness. Unfortunately, owing to the nature of the terrain and the heavy amount of equipment carried, there was one fatality.

BOAT TROOP

Boat Troop is responsible for all forms of water insertion methods, including diving and canoe work right down to swimming ashore on surfboards. In recent years, members of the SBS (Special Boat Squadron) have been stationed at Hereford and cross-training between the two units and several operations have been jointly carried out.

Boat Troop working with SBS during the Falklands War.

Boat Troop members are required to be proficient in handling Klepper canoes, Gemini inflatables, and other small raiding crafts. Infiltration techniques have even covered the firing of SAS soldiers from submarine torpedo tubes. The men are given a small breathing apparatus before being rammed three to a tube, one behind the other, in place of the torpedo. The inner hatch is then closed and the forward hatch opened, thus flooding the tube, at which time the men are pushed out into the open sea by compressed air.

MEDICS

SAS medics are highly trained personnel, undergoing an internal course at Hereford prior to attending one of several hospitals where they carry out practical work on real patients. The ability to carry out life-saving first aid and deal with most medical situations is of enormous value to a four-man patrol operating and cut off behind enemy lines. In many cases an injured soldier will have no prospect of skilled assistance and the patrol medic will be responsible for his wellbeing. SAS medics train to deal with everything from childbirth to massive gunshot wounds.

LINGUIST

The role of the linguist is vitally important when communicating with natives of an operational area. The ability to speak the mother tongue enables the patrol to develop a good relationship with the locals, especially during a "hearts and minds" campaign. It is also beneficial to have someone with a good working knowledge of the local language when interrogating prisoners of war.

Most SAS soldiers are actively encouraged to take a language course during their years at Hereford. Most courses are conducted at the Army School of Languages in Beaconsfield, although some languages are taught at Hereford. A language course is usually followed by a visit to the respective country, enabling the candidate to practice what he has been taught.

DEMOLITIONIST

The SAS Demolition Course is perhaps the most interesting of all individual SAS skills courses. This is owing to the vast variety of knowledge gained during the basic four-week course, to the extent that theory is backed up by practical demonstrations. Although there is a lot of classroom work involved, none of it is considered "boring." The practical work progresses slowly from the basic rules of handling explosives, to actually making the homemade devices. Many visits are planned during the course—places such as oil refineries, railway stations, telephone exchanges, and so on.

In any major conflict, demolitions will play a large part. Industrialized nations rely on producing war materials to sustain their frontline troops; destroying these installations will help defeat the enemy. The aim of the SAS Demolition Course is to do this using the minimum amount of effort. This means using complex formulas for cutting steel and placing explosive where it will do the most damage.

Because of the variety of tactical targets, many points of attack and placements of charges must be learned. Every target offers a different problem for the demolition team, as do the circumstances governing the attack. For example, in most cases the target will be heavily guarded. The SAS Demolitions Wing provides several different courses from teaching the basics to planning a full target attack.

SIGNALER

Communications are vital to any military unit; to the SAS it is a lifeline. All SAS operations require information to be passed between the patrol and headquarters unit, be it routine traffic or vital information about the enemy. Without communications there is no reporting, no casivac (casualty evacuation), no air strikes, and no extraction. The SAS signaler must learn to operate a wide variety of radios; likewise he must also learn the art of coding and decoding messages. The signaler normally shares his sleeping space with the patrol commander; this makes life easier as messages have to be sent at a set time, and it is normally quicker with two people. Additionally, the commander must tell the signaler what he wants to send and read any incoming messages.

SAS signaler, one of the hardest jobs within a patrol— but a vital one. He should not be smoking.

Morse code was originally used because its low frequency meant it could be hopped around the world. Then came the PRC 319 manufactured by Thorn EMI Electronics; this was a very powerful radio with a 50-watt output and an electronic message system. It was capable of data, voice, and CW transmissions over a very wide frequency band, 1.5 to 40 MHz. The PRC 319 also had burst transmission making it almost impossible for any enemy to intercept the traffic.

The SAS signalers now operate a compact man-portable satellite system (SATCOM). These have been around for some time and the very latest sets are only a little larger than the old PCR 319 radio. Satellite systems are extremely secure, offering communications from any point on the earth's surface directly back to Hereford.

Satellite communications systems include static, mobile, and portable units, the latter weighing less than 22kg (about 48 pounds), operating on UHF and SHF frequency bands. British forces currently use a number of Skynet 4 and 5 series satellites, which are in geosynchronous orbit, providing constant communications links worldwide. Once a transmitter/receiver has been set up and activated, it will locate and track a satellite with its dish antenna. Signals are transmitted on one frequency to the satellite which transfers it to another frequency via a transponder, boosting the signal and retransmitting it.

OPERATIONAL DEPLOYMENT

The current regular Special Forces unit is designated 22 SAS and is based in Credenhill. The camp itself is an old RAF catering station which was taken over and converted back in 1999. From here the SAS conducts their operations worldwide depending on the British Government's requirements. While the camp is vast, the actual number of SAS soldiers is only around 230 at best. These are divided into four Squadrons which are subdivided into four troops of roughly sixteen men. There is a permanent presence of SP Group which is in London and which liaises with the MoD and the other British Security units such as MI5. The camp is also shared with the Reconnaissance Regiment, a small unit of around one hundred personnel and which was derived from the old 14 Intelligence and Security unit that operated in Northern Ireland.

The SAS run an operational cycle whereby all SAS soldiers have the opportunity to move between standing operational commitments such as a tour in Afghanistan or serve overseas on specific training duties. Those who find themselves in the United Kingdom are normally either used in the antiterrorist team or put to training on new equipment or skills.

In recent years the SAS have been supported by the Special Forces Support Group, a new unit of around one thousand personnel which is situated in the old RAF base of St. Athens, South Wales.

SUMMARY

Selection and continuation is the most important part of any SAS soldier's training. It is a transition from being a regular British Army soldier to a member of the SAS. There are four main factors that govern this transition that stand out as extraordinary: fitness, self-determination, resistance to interrogation, jungle training.

There is absolutely no point in trying to gain access into the SAS if you are not fit enough—you will fail within the first few days. Additionally, there is no point in pushing yourself beyond what is required because you will need all your strength and endurance to pass selection.

If there is any one aspect that will help you prevail, it is self-determination. This single factor has to come from within you; it's what the SAS are looking for. They are not looking for six-foot muscle-bound hunks, but men who can complete the selection course and still be able to think and act at the end of each phase. In addition to your personal fitness you need to have good map reading knowledge and the use of GPS devices; knowing how to use the terrain is a major factor throughout SAS life.

When you knees hurt or you feel sick and want to stop, you must find the strength to go on. When the weight of your bergen is cutting into your shoulders and chafing your waist, you must learn to ignore it. When you're wet and cold with total exhaustion pressing on your body, you must not stop. If you can find the self-discipline to do all this, then not only will you pass SAS Selection, you will be a better person for the rest of your life.

The weather is bleak and the route seemingly endless—but the secret to being in the SAS is to never give up on Selection.

Interrogation training may sound horrific, but like everything else a SAS soldier must learn, it is a skill. Listen to what the instructors tell you beforehand and only give the interrogators your name, rank, number, and date of birth. You are advised to remove anything about your person that will help the interrogators such as clothing labels, tattoos, wedding rings—if not removed, they will use these against you. No matter what happens, just keep your mouth shut for twenty-four hours.

Jungle training is a place where much of your initial training will come together; if you can live and operate in the jungle you can operate anywhere. Come to terms with the environment; do not try to fight it as you will only lose. Learn to live with the constant discomfort of being permanently wet, eaten by insects, and making your way through vines that have teeth and claws or trees that have acid for sap. Above all, acquire the skill of living with other soldiers that rely on you. Train your senses, listen to the jungle, smell the jungles; come to love the jungle and trust your own judgment.

Chapter 3

BORNEO

As mentioned in the previous chapter, part of SAS Selection is jungle training. This derives from the early days of the modern SAS which originated from the Malayan Scouts where a four-man unit would patrol through the jungle. The jungle training provides the fundamentals of a four-man patrol working together as a team. The environment also demands good navigation while exercising stealth in movement and a disciplined routine, which is challenging on the individual and the patrol. These skills echo from the past, skills that allowed the SAS to roam through the jungle in small numbers undetected. While these skills may have been modified slightly over the years, the basic principles remain sound to this day. These skills are hard to describe, so the reader will forgive the lack of order set out below.

When the Labor Party won the elections in 1964, the new Prime Minister, Harold Wilson, and Dennis Healey, who was appointed Secretary of State for Defense, gave their full support to General Walker. Walker had been able to convince them that swift clandestine cross-border operations would help in the conflict against Indonesia. The operation was code-named "Claret" and was designed to preempt any buildup of Indonesian forces along the border of Malaysia. Neither Malaysia, Indonesia, nor Great Britain were officially at war, so the political risks were highly dangerous. However, Harold Wilson had promised and personally believed in supporting the new Commonwealth nations and was keen to suppress any threat from Indonesia.

President Sukarno of Indonesia vehemently opposed the move by Great Britain to unify the states of Malaysia, as he had his own designs on the rest of Borneo. The first sign of trouble had come as early as 1962, in the Sultanate of Brunei, when a small Indonesian-backed, anti-Malaysian element rebelled. British forces, however, quickly ended this revolt. In the same year, following the trend of recruitment, the SAS recruited and trained more local people, changing their name to Cross Border Scouts. To facilitate the mission, the SAS carefully selected forty Iban Dayaks, whose role was to carry out raids across the border into Kalimantan. The British believed the Ibans would be highly skilled at moving quietly and efficiently through the jungle, and in the event that a mission was unsuccessful, the Ibans would be the most likely to escape alive.

SAS use local Iban to help carry supplies in the jungles of Borneo.

Under the supervision of Major John Edwards of A Squadron, 22 SAS, training began in the summer of 1964. The first mission was in August, following which they were employed to be on guard along the border, especially in western Sarawak and around Bemban. So, when the Indonesian incursions started, A Squadron was already prepared to repel them. They knew the best ambush sites and helicopter landing zones and the "hearts and minds" campaign was working well.

Armed with the approval of both the Malaysian and British governments, General Walker started to conduct cross-border operations deep into Kalimantan. Initially, the maximum distance from the border was limited to 5,000 yards, but later in the same year it was increased to 20,000 yards for a small number of specific operations. All these operations were classified as top secret.

The force deployed to undertake operations comprised A and D Squadrons, 22 SAS Regiment and the Guards' Independent Parachute Squadron. The Guards were also

Claret Operations required that the SAS patrols used stealth and resourcefulness to take on superior numbers deep behind enemy lines. Moving by river was essential but also fraught with danger of ambush.

drafted in due to a chronic shortage in SAS personnel; therefore it was felt pertinent that they should undergo a course in SAS jungle tactics. Initially used for border surveillance, in September 1965 the Company, under Major L. G. S. Head, was allowed to carry out cross-border operations. They achieved excellent results, in one instance ambushing forty Indonesian soldiers, killing nine, and wounding many others. In 1965, several members who had served with the SAS in Borneo were later chosen to form the backbone of the newly formed SAS G Squadron. Other units used during operation "Claret" consisted of the Ghurkha Independent Parachute Company, 1 and 2 Squadrons Australian SASR, and detachments from 1st Ranger Squadron NZSAS. In addition, 1 and 2 Special Boat Sections, RM, also carried out small-scale raids on coastal targets on either flank.

"Claret" operations typically consisted of the interdiction of tracks, rivers, and other routes being used by Indonesian troops who were mainly well-trained regulars of the Indonesian National Army, the Tentera National Indonesia (TNI). These included para-commandos of the Resemeu Para Commando Angaton Daret (RPKAD) and marine commandos of the Korps Commando Operasi (KKO). Over twenty-two thousand TNI troops were deployed in the border areas, supported by large numbers of volunteer irregular forces.

In addition to the range of "Claret" operations, General Walker also laid down a set of regulations that governed the conduct of operations; they became known as the "Golden Rules." These stipulated that all operations had to be authorized by the Director of Borneo Operations personally, and that every operation had to be meticulously planned, rehearsed, and carried out with maximum security. Only tried-and-tested troops were permitted to be used and under no circumstances were troops to be captured, dead or alive, by the enemy. All operations were to be carried out with the aim of deterring and thwarting aggression by the Indonesians. Furthermore, the depth of penetration from the border was to be strictly controlled and it was emphasized that air support could not be given except in instances of extreme emergency. Despite the lack of air cover, artillery and mortar support was available. The artillery consisted of 5.5 in guns and 105 mm pack howitzers, while 81 mm mortars could be lifted forward by helicopter to the border region in support of operations.

The operational area had not been surveyed to any degree of accuracy other than from the air, thus detailed maps of the area were basically non-existent. The SAS patrol often left for their mission armed with only a green sheet of paper which may have shown the odd blue line indicating a river. Much of the mapping information was provided by the SAS patrols who would mark tracks and topographical detail after returning to base so the future patrols would benefit from their knowledge.

Much of the terrain detail and known tracks came from the local people, who still crossed the border freely into Kalimantan to trade their goods. They also provided valuable information on Indonesian troop movements. The SAS encouraged this and employed a policy of recruiting indigenous people, giving them title of Border Scouts. Unsuited to conventional combat roles, the Scouts were utilized primarily for information and intelligence gathering. In September 1963, a Scout post at Long Jawi, 28 miles from the border, was almost totally destroyed by a substantial group of well-equipped Indonesians. As a result, General

Walker restricted the role of the Scouts to one of intelligence gathering only. It was in this capacity that they really proved their worth to the SAS—frequently complementing individual SAS patrols by enabling communication between the forces and the native inhabitants.

From time to time patrols would deliberately enter local villages. They did this for two reasons: one was to make friends with the locals and the other was to elicit information on the enemy. They would approach with weapons lowered, and directly make for the headman or local chief in order to be respectful. This is where the linguist would come into his own, as he would be able to converse in their native tongue. After formal greetings, one of the first questions would be to see if anyone in the village was ill or had been injured; this allowed the medic to practice his skills, something the villagers saw as a sign of compassion. While the medic would treat men, women, and children (normally it was taboo to talk to or directly engage with the local women), the skill of winning the "hearts and minds" of the villagers provided vast amounts of information on enemy locations and habits.

In April 1965, Captain Robert Letts was put in charge of a patrol detailed with collecting intelligence on Indonesian communication routes across the border. When envisaging the typical mental and physical attributes of an SAS warrior, Captain Robert Letts might be considered to be an enigma. He was a great lover of music and literature, with a shy and self-effacing personality. Nevertheless, during the Borneo confrontation, this shy man exhibited great traits of bravery and initiative. The rest of his patrol consisted of Corporal "Taff" Springles, who, with the most experience, acted as both medic and signaler. Trooper Brown and Trooper "Pete" Hogg made up the rest of the patrol.

Navigation had been made easier for Letts and his men, as members of B Squadron had previously mapped out the swampy area in question. Nevertheless, it was still slow going as they were trying to keep their movements as quiet as possible. After six days of living in mud and water, the sound of an outboard motor drew them to a 12-foot wide stream, one of the main tributaries of the swamp, draining into the River Sentimo. Finding an area of dry land about 40 yards back from the stream, Letts and Hogg took off their bergens and left them with the other two men, intending to do a little reconnaissance. To lessen the likelihood of their tracks being spotted by the enemy, they took to the stream, walking and sometimes swimming along it. A small way downstream they saw a domestic water buffalo and heard voices, a sign that there was a village not far away. Letts realized that their progress through the river had been just a bit too easy: none of the usual debris of underwater logs or vegetation had blocked their way at any point. This could only mean that the waterway was regularly cleared to allow water traffic. The realization that an enemy-laden boat could come along without any prior warning was a great incentive for the men to move rapidly.

Back with Springles and Brown, an observation point (OP) was soon found, which was ideal for the purpose of monitoring traffic on the waterway. It was set on a loop in the stream where the visibility was clear 60 yards to the left and 30 yards to the right. Late in the afternoon their patience was rewarded as two boats, both containing two armed soldiers, paddled past them, heading downstream. At this point a signal was sent to base requesting permission to engage the enemy, should the opportunity recur.

Jungle ambush requires a quick and abrasive response.

The following morning no signal had been received from base granting the approval they sought—to engage the enemy (the rules for Operation Claret were very strict). Recognizing the opportunity was too good to pass up, Letts decided to use his own initiative if the opportunity to engage arose. The stream was obviously more important than it had previously appeared and probably served as a main supply route for men and equipment to the enemy base at Achan. The village that Letts and Hogg had discovered earlier possibly served as a staging point, which meant that any attack would bring swift retaliation.

A plan was drawn up, and at dawn, Letts positioned his men at various positions around the loop. At this time, the stream appeared difficult for boatmen to maneuver; with navigation needing more concentration the boatmen would be less alert to a threat of attack. The position of the loop also meant that once a boat had passed a certain point, it was no longer visible to those behind, thus creating the perfect conditions for an ambush. With Letts at the apex of the loop, Brown and Hogg to his left and Springles to his right, the men were ready to take on any boat, no matter what direction it came from. Letts also worked out an escape route; if they were successful and more boats appeared for them to ambush, they would need to get out of the area as quickly as possible.

They did not have long to wait. At 8:15 a.m. a boat came into view, but this time it carried three armed soldiers instead of two. The extra man sat in the stern, acting as sentry, alerting them to any dangers ahead. The next boat was the same: two armed men paddling sat in the front, and another, cradling a weapon, in the rear. This new situation upped the stakes.

As planned, the first boat floated past Brown. Unfortunately the second boat deviated from its route and crashed in the riverbank adjacent to his position. Due to the precarious nature of their situation, the crew failed to notice Brown or Hogg. As the boat struggled to regain its course, a third boat came into view; it also carried three armed soldiers. Eventually the first boat reached Springles' position and Letts gave the signal to open fire. He took aim at his first target—the armed sentry in the second boat—but was surprised when one of the paddlers grabbed for his gun with what seemed like amazing speed and lined him up in his sights. At this close range and in such a short space of time, Letts really had no chance to get out of the way and to change his aim to the bowman. He was in a perilous situation and in danger of losing his life, until Springles, who had spotted the situation, shot and killed the Indonesian soldier. By now there was a continued flurry of fire as all four members joined in. During the firefight, the crews of two of the boats managed to flip them over, throwing themselves in the water. Later, it was surmised that this was part of an anti-ambush drill, as some of them soon emerged on the bank with their weapons.

Observing the boat flipping over, Letts opened fire into the water hoping to inflict serious casualties. For the second time Letts found himself in grave danger as the remaining man alive in the boat had taken aim at him. The Captain realized his predicament—almost too late—and jumped to one side before turning his weapon on the man and firing. The Indonesian fell forward into the boat. For a split second, Letts looked away to see how Brown was coping with his boat when the man he had just shot seemingly came back from the dead. Although wounded, he raised himself up and took aim at Letts again. This time—luckily—Letts saw the action in time and shot him twice. Unlike previously, these wounds were definitely fatal.

With all this frenetic action going on, Letts had failed to notice that yet another Indonesian soldier had him in his sights. Once again Springles had spotted the danger and saved his life by shooting the man dead. At the same time another Indonesian soldier emerged from the water and made a grab for the dead man's weapon—he was killed alongside his companion.

Brown had dealt with the occupants of his boat relatively easily. They all lay dead. A fourth boat now came into view, but as the occupants realized what was happening, they quickly reached the bank and pulled their boat backwards, out of sight. Hogg fired a few rounds but the enemy was already gone. The firefight was over; the whole thing had taken only four minutes.

The boats, the riverbank, and the water were strewn with dead Indonesians with only one succeeding in escaping the ferocity of the patrol's attack. Letts gave the order to retreat and the patrol made haste to their OP and picked up their bergens before heading back in the direction of the border. With the Indonesians now alerted to their presence, they knew they had little time before the search parties were sent out. The hunters had now become the hunted and Letts decided that they would have far more chance if they relied on speed rather than concealment. Their extreme fitness served them well and they managed to cover a great deal of ground before the enemy mortared the ambush position. Apart from speed, the other advantage that Letts and his men had was that the Indonesians had no idea how big a force they were up against and seeing the damage done, probably thought the numbers were greater than they were. Therefore they would have expected such a large force to be moving more slowly through the jungle, giving Letts's patrol a much needed head start.

By the time dusk fell, the patrol had covered the majority of the nine kilometers (5.6 miles) needed to reach the border and Springles set up the radio to apprise base of the patrol's success. As luck would have it, there was an incoming message granting them permission to attack the enemy. Next morning the patrol cautiously made its way over the border and requested an extraction. An hour later they were winched up through a break in the forest canopy by helicopter. As they were being lifted, Pete Hogg got a vine caught around his neck and was almost strangled in mid-air.

Nevertheless, the whole operation had been a complete success and a few months later Letts was awarded the Military Cross for his part. Soon after, Letts joined the Australian SAS so that he could take part in the Vietnam War.

SAS weapons training in the jungles of Belize.

BELIZE

Most people associate the SAS and the jungle with the Malaysian and Borneo campaigns, but they have also operated in South America (in Belize, for example, which was threatened by neighboring Guatemala). Teams were also sent into Colombia to help in the fight against drugs.

The British Army retains a small garrison in Belize and occasionally the SAS have chosen to carry out their jungle training there. From time to time Guatemala has rattled its sabers and prepared to take Belize by force. Much of their rhetoric was taken with a pinch of salt, but on one occasion in 1972, they seriously intended to invade. The 1,800-strong garrison was rapidly reinforced by Britain.

Author's Note: This is normal in the SAS: one minute you were happy doing a spot of training in Hereford, the next you are winging your way over the Atlantic. On this occasion the C130 took off from Brize Norton, bound for Belize, with a refueling stop at Nassau in the Bahamas. Here we were given permission to leave the aircraft and stretch our legs while the aircraft was being serviced. It was at this stage that I noticed the RAF loadmaster beating one of the engines with a broom. *Funny*, I thought, until ten minutes later, when we were informed that the flight had gone "US" (unserviceable), and that we would have to spend the night in Nassau—I could have kissed him. If you have ever wondered why the sailors of old never returned from these islands, one breath of the relaxing air will explain everything. Forgetting the war, we were taken by a fleet of taxis to the Trust House Forte Hotel. Here the whole squadron was given grand rooms, some on the beach. "Eat what you like, gentlemen, for you happy hour has just begun." That was the manager's way of saying it was all free. Unfortunately, the next morning the aircraft was fixed and we continued our journey to Belize, most nursing a hangover.

As I was there, I will continue to tell this story in the first person.

We arrived about 2 p.m. and quickly set about equipping ourselves for the jungle. I was given a seven-man patrol. This was a bit heavy for the jungle, but it suited our task. We were to penetrate the thick jungle in central Belize, and move up to the border with Guatemala. By 4 p.m. we were ready to get choppered in when this guy arrived from the government.

He had come to tell us about the type of jungle in Belize. To a man of his stature, the entire squadron sat on the floor, as Major Rose introduced him. The first words he said were, "Fer-de-lance." We looked at each other in some amusement, *what the hell is he going on about?* He then proceeded to tell us all about this snake that could fly, it being extremely venomous and of the viper family. It would leap out of the trees and bite your neck and you would die. Great news. On top of this we still had to fight the Guatemalans.

We infiltrated at dusk and after clearing the LZ, found a spot to hole up for the night. To be fair, the boys went directly into jungle mode—no talking, no cutting—everyone alert. The next day we set off, and as two of the guys were new to the troop, I decided to carry out one or two contact drills. This we did, and I am confident that had we hit the enemy, the patrol would had held its ground and extracted in good order. Three days later, as we neared our objective, I suddenly saw the lead scout, Tony, move to one side and bring up his weapon. *Thump, thump*, followed by the cry, "Fer-de-lance." Initially thinking that this was an enemy contact, the boys behind me had gone directly into combat mode, ready to take on the worst. Two seconds later, as the word Fer-de-lance pierced the air, you couldn't see the patrol for dust. Fighting Guatemalans is one thing, flying snakes are another.

Our task was to observe a main jungle route that led from the Guatemalan side and stretched over the border into Belize. It was thought that the enemy might use this to infiltrate troops. In the event of an invasion, we were to report all enemy movements and strengths. If we were compromised, the plan was to bug out and run north to Mexico. Here an agreement had been reached with the British government to intern all its soldiers until the end of the war. Blood money was issued to make this journey possible; this time there was only American dollars as gold coins were not deemed necessary.

By day five, the RAF Harriers had arrived. They flew down the border in teams of four, stopping now and then to hover—threatening like giant flying insects waiting to sting. In full sight of any Guatemalan forces, they flaunted their massive firepower. It was enough; a day later they called off any thoughts of an invasion. Once more British "gunboat diplomacy" worked, and Belize was saved. We remained in the jungle for several more days carrying out "hearts and minds" before extracting to Belize and finally home. Before I left I purchased a T-shirt with "I fought the Guats" printed on the front and the words "ALL DAY" printed on the back.

COLOMBIA OPERATIONS

As part of a US/British effort to combat the huge narcotics industry in Colombia, members of the Regiment have been sent to Columbia since the '80s to train local police anti-narcotics commandos in infiltration skills. The police commandos are taught how to work in long-range patrols in hostile areas, destroy the factories where the drugs are produced, and to either kill or capture the criminals.

The SAS would teach basic jungle tactics, plus a few counter-terrorist methods. The training continued until the early '90s, normally with a full Squadron and a section of SBS. The Americans funded a large percentage of the operational equipment supplied to Colombia, but Britain also contributed night-sights, waterproof equipment, bergens, and military-type clothing. Despite what the media say concerning the role of the SAS in Colombia, the regiment never carried out any actual operations against the drugs cartels.

SUMMARY

When you arrive in the jungle for the first time it can be a little bit claustrophobic as the jungle actually surrounds you. The smell, noise, and the fact that the whole environment seems to be alive can be quite daunting. At this stage you must make up your mind whether you can operate in such an environment or if you're going to have problem adjusting. There is a saying within the SAS that "the jungle is neutral," and that is the best place to start.

One of the first things an SAS soldier must learn is how to live in the jungle—sleeping, movement, eating, sanitation, and a whole host of skills one would never think of. Making a bed for example—first you need to find a decent flat area with trees far enough apart to make your pole bed or hang your hammock. This is known as your *basha* space. There are two ways of improvising your bed in the jungle: a pole bed which is designed for long term and a hammock which is sewed when cutting (noise reduction is required), the latter being used when patrolling. These are all skills an SAS soldier must learn in order to operate and survive.

HAMMOCK

A hammock is a form of canvas bed derived from those used at sea by early sailors and adapted for use in the jungle. The modern military hammock can be slung between two trees in the traditional ship fashion or formed into a more permanent pole bed. The advantages of the hammock are manyfold, not least for lifting the soldier off the jungle floor, which although comfortable, is crawling with large-toothed insect life. Early hammocks were made from discarded parachute panels, which were folded and secured by the rigging lines. The modern military version is designed to be small, robust, and quick to erect. The hammock requires no cutting of trees, and is therefore quiet and used for single overnight stops during patrols. A shelter sheet to protect you from the rain is arranged over the hammock.

A "Pole Bed" will keep you off the jungle floor and provide shelter from the elements, plus a good night's sleep.

Pole Bed

A pole bed is a form of bed used mainly in the jungle to lift the soldier off the floor. The basis of the bed is two A-frames that when supported against a tree, have additional poles threaded through an SAS hammock and the ends placed over the A-frames. Opening the legs of the A-frame determines the height and security of the bed. A mosquito net and shelter sheet normally covers the bed. The pole bed is normally constructed in a secure base camp and used for long-term living in the jungle. When a pole bed is constructed and covered with a *basha* sheet, it forms a comfortable and waterproof dwelling.

Camp Routine

In a secure jungle environment, either the pole bed or the hammock will form your *basha* area—the place where you live. The longer you stay in one place, the better you can make it. For example, nothing moves in the jungle at night, therefore you have a few hours of darkness in which to dry out your boots and day clothing. If combat conditions are right, it is possible to remove your boots and socks and place them in a position under your shelter sheet overnight to dry out. During combat conditions it is also possible to cook a hot meal, but then you would move your location and find a secure location to hang your hammock and sleep.

Most SAS soldiers will make their *basha* space as comfortable as possible, and if in that location for a long time, will have a communal area where the evening meal is shared. Water is usually close at hand and a sanitation area for a toilet set aside for hygienic reasons. Life in this environment can be quite relaxed.

However, while patrolling during combat operations, the situation is completely different. Movement and talking are kept to a minimum, and strict rules are enforced. These include the allocation of emergency RVs in the event of an enemy ambush or the need to make a rapid retreat. Stops are made for signaling back to base, eating, and sleeping—all in different locations. Hammocks are put up silently and there is no smoking or torch lights allowed. In general, most men will wait for full darkness to change their wet clothes for the dry set in their bergen. This is known as "hard routine."

Jungle routine means making yourself as comfortable as possible, even if it is for only a few hours a night.

Equipment & Clothing

Jungle equipment consists of two sets of lightweight clothing—one which is used only for sleeping and is kept dry at all times in a watertight bag and the other which is used during the hours of daylight while patrolling. Before you finally get into your pole bed or hammock, you remove your daytime clothing (which is normally dripping wet from sweat or rain) and put on your dry night clothing. In the morning you simply change and put on your cold, damp day clothing. While patrolling you do this first and then put on your belt (in that order) so you are ready to move at a second's notice; after which you take down your hammock and stow your equipment away—then you make breakfast.

The belt kit is one of the most important items an SAS soldier has. Each man is responsible for constructing his own belt kit and packing the contents. The basic assembly will include a belt, which can be anything from a heavy-drop quick-release strap to a standard issue belt which has been adapted to hold the pouches and sometimes a supporting yoke. There are no hard and fast rules, but the belt must carry the following: two ammunition pouches, each capable of holding four magazines; two water bottles, one with a metal brewing mug; two large-sized mess tin pouches, one for permanent use: survival kit, emergency rations, flares, medical kit etc.; and one for daily use: mess tins, and consumable rations for that day. The belt should also have a knife or Golok, and several "first field" dressings attached to it.

"Grab your belt kit" is a common phrase within the SAS, and this sums up its importance. Even while at the Stirling Lines military base, the belt kit, with the exclusion of ammunition, always remains ready to go at a moment's notice. In the field, unless sleeping, it remains around the soldier's waist at all times. During an operational contact, the bergens may be dropped, but the belt kit remains with the soldier to the bitter end. The unwritten rule is that your belt kit and weapon are never more than arm's reach away.

Weapons

Weapons for jungle fighting need to be short range, and in the early days, many of the SAS soldiers preferred to use shotguns. However, weapons will depend on the mission. For example, if it's just reconnaissance, then there will be no need for heavy weapons as the patrol will want to remain hidden. If there is a greater need, as in the "Claret" type operation carried out in cross-border raids into Indonesia, then the SAS would carry a great deal of firepower. Today a four- or six-man patrol will normally carry the Demarco, but again it will depend on the type of operation and what is needed.

Jungle Methods of Entry

It is not easy to access the jungle, mainly owing to the lack of roads and the restrictive nature of the terrain itself. Several methods of entry have been tried and tested over the years, the first being by parachute which was known as "tree jumping." This technique for parachuting into the jungle was devised by Captain Johnny Cooper and Alistair MacGregor of the Malayan Scouts. The idea was to land on the high jungle canopy, which would entangle with the parachute canopy, whereupon the men could release themselves and rappel down to the ground. Hence this became known as "tree jumping." The first true operational jump took place in February 1959, when fifty-four men tree jumped into the Malayan jungle to assist troops on the ground. Hailed as a success, the technique became accepted as a standard tactic. However, it soon became apparent that this was more dangerous than it first appeared—three men had lost their lives tree jumping in Operation Stone alone, and by the end of the Malayan campaign the practice was discontinued, aided by the increasing use of helicopters for jungle missions.

While helicopters were a great asset, they were not always easy to set down; rappelling from helicopters was not common at the time, and so a clear landing site had to be found. On occasion helicopters would use a winch to extract a patrol. Once the SAS patrol was on the ground, it was much easier as the men could select a location and clear it sufficiently for the helicopter to land. In an emergency where a medevac was needed, the demolitions man would sometimes blow down the trees in order to construct a suitable LZ.

In some cases it was possible to use boats as the rivers are a more traditional method of transport with the local

Jungle Tactics involve crossing any large tracks with extreme care.

population. However, in the case of the SAS, movement by boat did increase the chances of ambush by the enemy.

Jungle Patrol Tactics & Skills

Patrol skills include movement, camouflage, ambush and anti-ambush drills, contact drills, and RV procedures, all practiced to ensure that the SAS four-man patrol can operate undetected in a hostile environment. Most SAS soldiers learn their basic patrolling skills in their original unit, especially if they come from an infantry regiment, but SAS patrol skills are taught during the jungle training phase.

As most SAS patrols are made up of only four men, the commander must utilize his men to the fullest, placing each man within the patrol based on ability and individual skills. For example, he will not make the signaler or medic act as the lead scout; rather he will choose the demolitionist or linguist. Luckily, most SAS soldiers will have more than one skill, which allows the commander greater flexibility in the structure of his patrol.

While all SAS personnel learn contact and ambush drills and RV procedures during continuation training, no two theatres of operation are the same, and a good patrol commander will insist on rehearsals prior to insertion. Likewise, once inserted, the patrol must be disciplined by hard routine, obeying SOPs to avoid detection by the enemy.

Patrolling through the jungle is extremely hard. Generally, there is no cutting, which means the patrol must push their way through the jungle; no noise means they must do this as silently as possible. If you add to this the fact that SAS patrols are taught to avoid jungle tracks, ridges and valleys, and contour to steep-sided jungle buckets (SAS name for large hill), you can visualize how hard movement could be. It was not uncommon for an SAS patrol to cover less than 1,000 meters in a day.

While doing this, the lead scout would need to read any sign he might come across, as well as signal rest stops and indications of enemy presence, the latter being done through hand signals. From time to time the patrol commander might indicate a prominent spot as an Emergency RV should they get ambushed; or signal ambush positions if they came across a track that looked like it was used by the enemy. Other drills practiced would cover "shoot and scoot": this is when the patrol has run directly into the enemy at very close quarters.

"Shoot and Scoot" is a standard operating procedure (SOP) used by the SAS, particularly in jungle warfare. Conceived by Lieutenant-Colonel John Woodhouse, it was designed to prevent heavy casualties during a surprise encounter with a hostile force when there was no point in holding the position. It works as follows: the SAS soldiers, when attacked, would reply with a heavy barrage of fire-power before quickly making their individual ways to a pre-determined emergency rendezvous point. In a small patrol of four men, once the lead scout has safely withdrawn, all members will rally to the patrol commander.

Chapter 4

OMAN

The SAS have also taken part in what could be described as traditional warfare, as they acted more in an infantry role, albeit a specialized infantry role. The war in Oman between 1969 and 1975 is one good example and possibly the Falklands War is another. In both cases they operated alongside regular troops. In such a role, patrol tactics and raids were very much a part of daily life.

OMAN WAR

The war in Oman stands out as a classic SAS operation. It was tailor-made for the SAS, as it would test their skills as never before. It was also a war the SAS desperately needed. The period of relative peace that had existed since the end of the troubles in Aden had depleted the regiment's ranks. Despite the extensive training programs embarked upon by the SAS Squadrons, many of the younger members started to leave, tempted at the time to be highly paid bodyguards or by more active security jobs. Unfortunately, this minor exodus took many of the best corporals and sergeants; however, some did return once the Oman War got underway.

The war was fought in secret, but its importance—both militarily and politically—had far-reaching effects over the entire Middle East. Moreover, its success helped stabilize the attitude of many neighboring countries, pushing back the tide of communism that threatened to engulf the rich oil fields on which the West depended. It offered a challenge that the regiment eagerly accepted and turned out to be a classic counter-insurgency campaign of modern times.

In size the war was small, yet it combined all the elements of modern warfare. Navy, air force, and army were all united with one goal: to win. It cannot be said that the SAS won this war on their own; much of the fighting was done by the Sultan's own country-men and army. Yet, undeniably, the one thing the SAS did do was bond

The Firquat worked in a similar manner as the Ian tribesmen in Borneo during the Claret Operations. At times they could only be recognized by the weapons they carried.

together the Firquats. These Firquats and SAS went on to become the lead elements of most battles in the early days and it was a mixture of trust, hearts, and minds that won the Dhofar War.

FIRQUAT

The name Firquat is a rough Arabic translation for a military company and was used to describe the irregular Dhofari troops during the Oman War. Many were from the southern coastal villages, while others had previously served as anti-government guerrillas and had been granted amnesty by the Sultan. These men were raised and trained by the SAS and formed into Firquat companies.

The idea of Firquat troops was first put forward in 1971, after a meeting between Major Tony Jeapes of 22 SAS and Salim Mubarak, a former leading figure within the Dhofar Liberation Front. Mubarak suggested that an anti-guerrilla fighting force could be formed using disgruntled Dhofari tribesmen; he also suggested a name —Firquat Salahadin (Company of Salahadin). Jeapes agreed and soon, with the cooperation of local tribal elders, the SAS were training Firquats at Mirbat. Although they took part in battles, the Firquat were considered to be reconnaissance rather than infantry troops. Their task was to gather intelligence, usually through village gossip, and to pass it on to the British troops.

Most of the war was to take place in the south, on a huge mountain known as the Jebel Massive. It was a strange refuge for wild tribes and freedom fighters. In the summer it was a place of great beauty, where lush, green grass flowed in the cooling winds, trees gave homes to birds and other small animals, and on rare occasions larger game could be seen.

The whole Jebel Massive had been taken over by the rebels (known as the "Adoo"). Only the coastal towns of Salalah, Taqa, and Mirbat remained free, and even

Jebel Massive Southern Oman where the SAS fought for six years; despite the war it remains a place a great beauty.

these towns had been infiltrated by the forces of the Adoo. The rebels roamed free over the Jebel Massive, eating further into the beleaguered areas around the southern capital of Salalah. The British RAF base just north of Salalah was itself virtually under siege, and the Sultan's small defense force was fighting a losing battle.

Although not a military man, the old Sultan had decided to send his only son, Qaboos, to the Royal Military Academy at Sandhurst, where he became commissioned into a British Regiment. His stay in England was far from wasted; with skill and the encouragement of his friends he had observed the workings of various councils and committees, and in general had familiarized himself with the workings of a modern state. His return home had not been a joyous one. The young Sultan could see the plight of his country and argued for change. His father's answer had been to further restrict his son's movements and accuse him of becoming a "Westerner." However, the young Qaboos bided his time. His chance came on July 23, 1970, when a palace coup d'etat took place. Qaboos, aided by the young Sheikh Baraik Bin Hamood, deposed his father in a bloodless takeover. Within weeks, soldiers of the British SAS had been sent to provide advice and assistance.

For the SAS, the first action started in the northern Oman peninsula, at the very entrance to the Straits of Hormuz.

The author takes up the story:
A Squadron member was getting married and most of us were attending the ceremony. So around midday, dressed in our finest suits, we prepared to leave the barracks when a loud voice called out "Get your kit, we are going."

"Where to?" Someone shouted at his receding back.

"Middle East" was all I heard before I, too, broke into a run, heading directly for the accommodation (*basha*). There were a few guys still getting dressed for the wedding, but the bulk had already left for a few beers before the ceremony started. The sudden appearance of the Squadron Commander soon made me realize that this was not a drill. He quickly informed those present of the latest situation. It would seem that some British Intelligence guy who had been lying low in a village off the Musandan Peninsula had picked up information that a huge arms shipment from China was due to arrive at the tiny coastal village of Jumla.

This arms shipment was to be accompanied by a group of Iraqis with communist sympathies. According to British Intelligence, this was translated as the main communist command center moving into the area to take advantage of the turmoil within Oman—the SAS were going to snatch them and nip the whole thing in the bud.

> **Author's Note:** All SAS soldiers are permanently packed and ready to go at a moment's notice. This usually means having a bergen (rucksack) packed with all the essentials needed for any operation. This includes clothing and standard belt equipment. Additional items such as body armor, night vision equipment, etc. can be added at the last minute. Squadron weapon rolls are always ready and collected from the armory as transport arrives.

A few hours later, with most of the squadron still dressed in suits, we made our way to RAF Brize Norton and some hours later landed at the British airbase in Sharjah. Without leaving the confines of the airport, we loaded our bergens and weapons onto several trucks and were driven deep into the desert. For all intents and purposes this was done for security; in reality the whole Squadron was dumped in the middle of nowhere. Yes still dressed in our suits! Like all good SAS men, we dived into our bergens and changed into combat gear before putting up basha sheets for shelter and making the inevitable brew of tea. Later that evening, several choppers came in and we were ferried over to a camp site on the eastern coast. The next day a Royal Navy minesweeper, complete with several rigid raiders manned by the SBS, arrived off our coastal campsite—and thus the plan unfolded.

Bukha from the sea, little has changed.

It would seem the SAS would be inserted by the SBS onto the small beaches both north and south of Ghumda village where the Chinese ship would dock. This would be done during the hours of darkness and allow enough time for us to scale the high rocky peaks that towered out of the sea and surrounded the entire village. By dawn, with the SAS in place to stop any enemy running away, a combined force of the local Arab States using loaned American landing craft would assault the village and capture both the Iraqis and the Chinese equipment. On the surface it seemed to be a sound plan. Not only would it protect the Straits of Hormuz, through which half the world's oil passes, but it would also stop a major political shift in the area. But this is what really happened.

The night before we left our training area to board the minesweeper, a priority signal arrived. By the light of our campfire the Squadron Commander read it to us. He reported that heavy machine guns and mortars had been unloaded, together with several large boxes of mines. It was suspected that the enemy had received wind of our assault and planned to oppose it. As God as my witness, the final words of the message were: expect to take up to 50 percent casualties. Not the kind of news that inspires you to fight!

As if being blown to smithereens on the beach wasn't enough, the whole plan was delayed for twenty-four hours due to bad weather. The next day, as they loaded our sea-sick bodies into the rigid raiders, I would have happily taken on the enemy single-handed rather than stay a second longer on that ship. As we approached there was nothing to see other than a few faint lights coming from the village— then we hit the beach at full speed. I was actually thrown from the raider in what looked like a spectacular, headlong dive. Somehow I landed in a perfect para-role and came up on my feet. Several of the guys said the dive looked real gung-ho—I just wanted to kill the SBS boat handler.

Author's Note: When it comes to insertion from the sea, and while Boat Troop is more than capable, the SBS often take on this role. They are better equipped to insert SAS troops.

Almost instantly we started to climb. It was steep and hard going, but Mountain Troop led the way, choosing the easiest route. An hour before dawn we were in position, overlooking the village to the open sea beyond, our only casualty—a broken finger.

TROOP SKILL:

This was a perfect example of how having specialist troop skills can aid an operation. In this case G Squadron Mountain Troop selected a route that was easy enough for men with heavy bergens and weapons to climb in total darkness and in secrecy.

As the light rapidly improved, we spotted the American landing craft about half a mile out to sea. They were heading at top speed for the beach—I have to say that from our vantage point, high above the village, it looked impressive. The landing craft hit the beach, and several shots sounded out as excited Arab soldiers fanned out across the beach heading for the village. Slowly the villagers started to emerge from their homes to greet the soldiers (strangers and visitors were rare in this remote area); tea and coffee were distributed and everything seemed very amiable. Then one of the white officers got tough. "OK. Where are the Communist rebels? We know they are here in Ghumda."

This is the actual cliff face to the side of Bukha that had been climbed in total darkness led by G Squadron Mountain Troop.

"If they are in Ghumda, why are you here in Bukha?" was the surprised reply. Yes, we had invaded the wrong village. There were no more than half a dozen villages along this hostile coastline, but we picked the wrong one. By the time the head-shed had confirmed the fact and sent us racing across the mountain top, it was too late. The rebels had gone, and so had most of their equipment.

We were all pulled back to the British base at Sharjah, where we sat around while a new plan was hatched.

Apparently, some of the enemy had taken refuge in a stronghold called the Wadi Rawdah. Even my best effort at explaining to anyone what his place looked like would not do it justice. It was a mighty bowl within the Jebel structure. The sheer, rocky walls towered a thousand feet even at the shortest place. On the seaward side there was a natural, narrow, split in the rock structure, which allowed for entry and exit. The whole valley was home to a strange tribe called the Bani Shihoo, reportedly, a wild and vicious people who had rarely seen a white man. Their main weapon, apart from the odd musket, was a small but vicious-looking axe.

The operation restarted with a parachute night drop carried out by two free-fall troops. I think I am right in saying it

was to be the first operational free-fall drop the SAS regiment had ever undertaken. The rest of us would go in by chopper as soon as the boys had secured an LZ. The aircraft took off around at 3 a.m., and an hour later the men jumped from a height of 11,000 feet. Paul "Rip" Reddy was killed. Of all the deaths in the regiment, this one stunned me. To this day I can recall talking to him minutes before he left. He was so young, fit, and full of life, but that would sum up most men who have died in the SAS.

Rip Reddy was killed during the first SAS free-fall operation into the Wadi Rawdah. This is the helicopter that arrived with support troops and to collect his body.

Most of us were still sleeping when the call came telling us to grab our kit and get on the chopper, while also informing us that Paul Reddy was dead. The chopper had a machine gun in the open doorway but we all tried to get a glimpse of the Wadi Rawdah—from the air it takes your breath away, like being on another planet. As the chopper touched down we all jumped clear, fanning out in a defensive arch. That's when I saw two of free-fall troop bring a body bag forward and place it on the chopper. Paul Reddy was a good man.

We stayed in the area for about two months, using Sharjah as a base. I was pulled off to do Psychological Operations, an interesting task that gave me a good insight into how a propaganda machine could help win wars. This lasted a month or so and eventually we went home . . . but only for a short while.

PsyOps

While not a widespread skill within the SAS, they have used it to great advantage in the past. In the case of Oman, it meant setting up a local radio station (SAS Int Sec controlled) to inform the scattered population of the new Sultan taking over from his father. My job was to actually go into all the remote villages and distribute radios so they could listen to the news. We also made T-shirts for the children on which we printed the Omani flag. The Sultan, in a quest to get himself accepted, traveled around all the villages; we would go in the day before and warn off the village elders and dress all the children in nice new T-shirts, hence the visit was a great success. A lot of anti-communist literature was also produced and air-dropped over the enemy-held villages; many of the local population surrendered once they knew things would improve under the new Sultan. One thing I learned during my brief PsyOps employment was to always tell the truth, even if it was against you— people like honesty.

Locals holding up PhyOps pamphlets dropped by the SAS during the Oman War.

After the abortive raid in the north, the war in Oman continued, but this time in the south. After the first initial recess by Colonel Watts and Major De La Billiere, it was decided that we were to occupy the southern coastal towns. The SAS HQ was established at Um al Gwarif, on the outskirts of the southern capital Salalah, a few minutes from the RAF base. While Salalah and the air base took several long-range attacks, both were defended by an outer ring of firebases called "hedgehogs." These were situated in a defensive arch between the Jebel Massive and the air base and were manned by the RAF Regiment. They were well-equipped with mortars and "Green Archer" (an enemy fire tracking system). This ensured that the Adoo could not bring their heavy weapons in too close. The more distant towns of Taqa and Mirbat were open to the full brunt of the Adoo—hence these locations were manned by the SAS. In the days before the war was taken onto the Jebel massive, these locations came under constant attacks. Hardly a night would go by when either Taqa or Mirbat did not come under fire. Life for the SAS training teams, prior to taking the war up onto the Jebel, was confined to village life raising and training the Firqats.

BATT (British Army Training Team)

BATT is the name attributed to the first SAS units to enter the war in Oman. The war was kept secret from the general public, hence the designation British Army Training Team (or BATT, as it was commonly known). SAS detachments were first sent to Oman in July 1970, directly after the Sultan Qaboos had just deposed his father in a coup. The teams occupied the larger villages along the south coast of Oman and here they raised, trained, equipped, and lead a militia army.

The Firquat mustered around eighty men, and they were fearsome fighters. During the old Sultan's frugal regime, many young men had left their villages and traveled widely within the Arab world. Fortunately for the SAS, some had taken training with the Trucial Oman Scouts and had very good knowledge of British military tactics. It was common to see the section commanders lead off giving the same hand signals that could be seen on the training areas of Great Britain.

I would be amiss not to stress the importance of the bond between the Firqat and the SAS. Sure they had been trained to act as a military body, but in reality they were far from it. Yet they possessed a feeling for their own backyard that the SAS did not have. It was not uncommon for them to wander into battle with their rifles slung over their shoulder; then quite suddenly they would drop to the ground and start darting forward. It was a movement the SAS soldiers came to recognize, it meant "Adoo." In battle they were courageous, always dashing into the fight even if sometimes their firing became a little erratic. They were also honest; if for some reason the SAS unit did not do as they requested, they would soon make their point obvious. At the same time, when they were around, one could guarantee a good night's sleep. Studying the Firqat gave one some idea as to what the Adoo were like. In fact, as the war progressed, many of the Adoo captured or surrendered would help swell the ranks of the local Firquat. Captured Adoo differed greatly. It was not uncommon to find after the battle, dead or wounded, Adoo dressed in a better uniform than the SAS. Khaki Shorts and shirt, ammunitions belt, water and an AK-47; this would all be topped off with a blue beret complete with red star and a copy of the *Thoughts of Chairman Mau"* in his pocket. Dead Adoo were often stripped of their weapons and ammunition by the Firquat.

The SAS were known as the British Army Training Team. This picture is a typical example of how an SAS troop looked. Note the early use of the M16 and the "TOKI" walkie-talkie radios for local communication.

The bulk of the coastal villages were sheltered under a small escarpment and protected by an old "Beau Gest"-type fort. A troop of Baluchistan soldiers normally occupied the fort, which in turn was surrounded by a razor wire fence in a half-moon shape. SAS always built a mortar pit and some had an artillery piece which would help support the village when it came under attack—which happened to be most days. The attacks were mainly standoffs that lasted no more than

twenty minutes and normally consisted of mortar and light machine-gun fire. They became so routine that the SAS teams would be waiting—trying to anticipate from which direction the attack would come. SAS mortar teams would sit with two bombs at the ready; as soon as the *plop plop* of the enemy mortars was heard, the SAS would dispatch retaliatory rounds immediately. The exchange was normally followed by full small-arms fire. In addition, the SAS would do night penetration patrols and lay unmanned demolition ambushes.

Operation Jaguar was launched to establish a firm base on the Jebel, beginning in October 1971. Dispensation had been given by the senior Qadi (religious leader) to all Arabs fighting during the Ramadan period, which is normally a time of fasting. This was the start of the war proper: almost two full Squadrons of SAS, together with their Firqats, spearheaded the operation. Additionally, several companies of the Sultan's Armed Forces (SAF) and various support units also took part. The whole force was led by Colonel Johnny Watts, a brilliant commander. Watts had a quick, decisive mind, yet he would not commit the SAS without committing himself. He was also no stranger to the battlefront and could often be seen running forward with one gun group or another carrying boxes of ammunition, shouting orders as he went.

The Shorts "Skyvan" of the Omani Air Force was the ideal air vehicle to resupply and for transportation during the whole Oman War.

Getting the men onto the Jebel was not as easy as one

would have anticipated. A diversionary plan had been in place for several weeks. (Heavy patrolling had been initiated from Taqa and Mirbat and directed at the Wadi-Darbat. The Darbat had always been an Adoo stronghold and the intention was to make it look as if a full-scale attack were imminent.) Helicopter hours were limited, and although some could be used in the initial lift onto the Jebel, most would be required to ferry ammunition, water, and rations to support the early days of the operation. They would need to sustain the war effort until an airstrip could be built and secured that would allow the lager Skyvans in.

While SAS and Firqat from Mirbat and Sudh climbed the Jebel from the east by a feature known as Eagles Nest and worked their way westward during the day, this also averted the Adoo's attention. Meanwhile, a full Squadron undertook a grueling march to occupy an old airstrip at a place called Lympne. It still stands out in the minds of those men who did the march. The route was over very difficult terrain; the bergens each man carried contained enough ammunition and water to last for several days. But the march and climb were so severe that, upon arrival, even the fittest SAS were totally exhausted. Luckily the Adoo were occupied elsewhere and later that morning SOAF helicopters and Skyvans started bringing in the second SAS Squadron.

Colonel Watts decided to move to a more defensible position. So on the morning of the second day, the SAS and Firquat units started to move westward. With the Firqat in the lead, Mountain Troop of G Squadron topped the small rise around an area called Jibjat when they came face-to-face with a large Adoo group having breakfast. A firefight rapidly developed and the surprised Adoo started to break-up, trying to disperse. We were so close that everyone started to run with both Firqat and SAS going into a full frontal attack. They overran the Adoo position and continued to clear the area to the south, whereupon any further advance was stopped by a large wadi (ravine). The firefight continued to rage, supported by the SAS heavy gun teams which had joined in the fight. GPMGs were adapted to the heavy support role and consisted of a three-man team. One would carry the whole unit, gun complete with tripod, while the other two would carry ammunition and act as gun loader and spotter. In the early days of the war, these gun groups proved decisive in the winning of the firefights; initially the SOAF Strickmasters jets were not so quick on the scene. For the next two days, small battles could be heard flaring-up at one location or another. By day three, Colonel Watts had split his force into three battle groups, two of which were dispatched to clear the Wadi Darbat and a ridgeline known as the gatn (pronounced "Cuttin" by the locals).

For several days the Adoo fought with everything they had, mistakenly thinking that this was nothing more than a short operation by the Sultan's Armed forces (SAF) and that in a few days they would give up and leave. By October 9, the initiative was clearly on the side of the SAF and the Adoo broke up into smaller groups and disappeared in the small bush-covered wadis (deep valleys).

SAS life in in the village of Taqa—the flat-roofed house immediately to the right of the old tree was known as the BATT house. Slaves were sold around the tree in the middle which acted as the local market area.

As both sides eased back a little, a main base was established at a place known as "White City." By this time the choppers were quickly running out of flying hours and desperately needed servicing. Likewise, ammunition supplies were dwindling—especially mortar bombs—and water was at a premium. It was Colonel Watts's suggestion to construct an airstrip in the middle of White City so that the Skyvans could alleviate the supply situation. As troops began arriving into the location, the Firqat were sent to picket the high ground while SAS men set to work constructing the airstrip. They worked all night, several times coming under heavy enemy fire, but by dawn they were ready to receive the first aircraft. Again the battle flared up. Each time a Skyvan landed, the Adoo were waiting; mortar bombs began to rain down and small arms attempted to shoot several aircraft down. To prevent this, heavily armed dawn patrols were sent out to engage and occupy the enemy while the aircraft unloaded. This period saw some of the heaviest fighting of the war.

There are many untold stories of the Oman War, the most outstanding being an attack by the Adoo on the coastal town of Mirbat. At dawn on July 19,1972, a large rebel force attacked the Port of Mirbat in the Dhofar Province of Southern Oman. This was the last great attack by the Adoo in the Oman War and only by a single stroke of bad luck had they miscalculated. There were two SAS Squadrons in Oman at the time, carrying out an end-of-tour hand over.

At around 5 a.m., the picket at the top of Jebel Ali, a small hill 1,000 meters to the north of Mirbat and halfway towards the Jebel Massive, was being manned by a section of DG (Dhofar Gendarmerie). Jebel Ali provided a dominating feature that protected the town and the surrounding coastal area. The DG were the first to be killed. Stealthily at first, cutting the throats of those still asleep, and then as the alarm was raised, the Adoo opened up with small arms. The BATT house heard this exchange of fire. The commander, Captain Mike Kealy, and his men observed the fire coming from Jebel Ali and the waves of men advancing towards the old Beau Geste type fort.

The battle for Mirbat will live forever in SAS history and the old fort at Mirbat still stands to this day

Captain Kealy shouted orders for the 81mm mortar to open fire in support of the Jebel Ali, while the rest of the SAS took up their positions behind the sandbagged emplacements and awaited confirmed targets. As a safeguard, the signaler was ordered to establish communications with SAS Headquarters at Um al Gwarif. Additionally, the SAS Fijian soldier, Labalaba, left the house and ran the five hundred meters to the DG fort, where he manned an old World War II twenty-five pound artillery piece. By 5.30 a.m, there was enough light for Kealy to make out the silhouette of the gun position and fort. Suddenly, a vast amount of small arms fire started pouring into the town. Through the mist, figures could be seen approaching the perimeter wire from the direction of Jebel Ali—the Adoo were attacking in waves. As the battle was joined, both SAS machine-gun bunkers opened up; at the same time the 81mm mortar increased its support to the battle. In the gun-pit by the DG fort, the twenty-five pounder sent shell after shell into the enemy at point-blank range. The battle flowed back and forth until a radio message came through on the Tokki (a small commercial walkie-talkie used by the SAS through-out the Oman War) from Labalaba, stating that he had been hit in the chin while operating the twenty-five pounder. A man of such stature is not given to reporting such trivia, and those at the BATT house suspected he was badly injured.

Captain Kealy immediately sent Labalaba's countryman, Takavesi, to his aid. The gunners provided supporting fire from the roof of the BATT house as they watched Takavesi run the gauntlet of tracer bullets and exploding shells, div-ing headlong into the gun-pit. He found Labalaba firing the big gun on his own. At that time, Labalaba gave no indica-tion that he himself was injured. Instead he indicated the unopened ammunition boxes and the desperate need to keep the gun firing. Much of the Adoo attack was now directed against the gun. Takavesi left the pit to solicit help from the DG fort a few meters away. As the rounds zipped past his head, he banged on the fort door until, at last, he was heard. The first man to appear was an Omani gunner and Takavesi grabbed him. Together both men raced the few meters back to the gun-pit; as Takavesi cleared the sandbags, the Omani gunner fell forward as a bullet hit him in the stomach.

It was now light enough to see groups of men near the outer perimeter fence that covered the three open sides of the town. Behind them, wave after wave could be seen advancing towards Mirbat in support. Abruptly, several rockets slammed into the DG fort, causing great chunks of masonry to be blown from the ancient wall. From the BATT house roof a new threat could be seen: the Adoo had breached the perimeter wire. Swarms of young men threw themselves headlong onto the barbed, razor-like steel; as one died another took his place. Men could be seen all along the wire, but the main breakthrough seemed to be in front of the fort. Once inside the perimeter, the Adoo advanced on the fort and gun position in large numbers. The gun, now leveled directly at the wire, fired at point-blank range into the charging figures. Suddenly, Takavesi cried out, "I'm hit," and he slumped back against the sandbags . . . but grabbing his SLR, he continued to fire. Labalaba made a quick grab for a small 60mm mortar that lay close by. He almost made it; then a bullet took him in the neck and he fell dead.

Communication with the gun-pit had been lost and Captain Kealy decided that he and an SAS medic, Trooper Tobin, would risk going forward to give assistance. Before they left, Kealy contacted Um al Quarif, informing them that air cover was desperately needed. He also requested a chopper to take out Labalaba. Additionally, if the firefight continued at its present rate, additional ammunition would be required.

At the SAS HQ in Um al Quarif, a relief force was quickly assembled from the recently arrived G Squadron; they were already dressed and equipped as they planned to spend the morning test firing weapons. It took about five minutes for twenty-two of them, under the command of Captain Alastair Morrison, to get together an impressive array of weapons. Eight GPMGs and several grenade launchers were among them.

Back in Mirbat, Captain Kealy and Trooper Tobin worked their way forward to the gun-pit; as they approached, the fighting increased and both ran for cover. With a final dive, Tobin rolled into the gun-pit; Kealy was about to follow, but realizing there was not enough room and tripping over the dead body of a gendarme, he threw himself headlong into the sandbagged ammunition bay.

Trooper Tobin could not believe the mess. Labalaba laid facedown and very still, Takavesi sat propped against the sandbags, weapon still in hand. Nearby the Omani gunner moaned, clutching the wound in his stomach. Assessing the priorities, Tobin quickly set up a drip on the seriously wounded Omani gunner. Takavesi was severely hit in the back, but despite the loss of blood he continued to fight, covering the left side of the fort. The firefight reached its height and the Adoo made a real effort to overrun the gun. As Kealy concentrated amid the mayhem, he saw Adoo close by the fort wall; several grenades were thrown and bounced by the lip of the gun-pit before exploding. An Adoo appeared at the side of the gun-pit but Kealy cut him down.

In the pit, Tobin reached over the inert body of Labalaba. Realizing there was little he could do, he made to move away when a bullet hit him in the face; he fell by the side of the big Fijian, mortally wounded. The gun-pit seemed done for when suddenly, there came an almighty explosion: the SOAF jets had arrived.

Despite the monsoon weather, the pilots had dived out of the cloud just feet above the ground. Firing heavy cannons, the first two jets made pass after pass, driving the Adoo back into a large wadi outside the perimeter wire before finally dropping a large 500-pound bomb where the Adoo had taken refuge.

By now the G Squadron relief force was airborne on three choppers and was rapidly heading down the coast. They covered the 30 miles to Mirbat in about ten minutes. Owing to the low cloud cover, they were dropped off to the south of Mirbat and instantly made contact with an Adoo patrol that was covering the rear. The Adoo, consisting of one older soldier and three youths, were held up in a cave and refused to surrender. Several 66mm Law rockets slammed into the entrance, followed by abusive fire from several GPMGs. The Adoo picket was quickly neutralized.

The actual artillery gun fired by Labalaba at the battle of Mirbat is now a military museum piece. Labalaba was a real SAS hero in every sense of the word.

With the jets taking the sting out of the Adoo, Kealy had time to crawl forward and examine the gun-pit. He could see that the Omani gunner was still alive and so was Tobin, although his wound looked horrific. Takavesi lay listless against the sandbags; his whole body seemed covered with blood—but he smiled. Then the SAS relief force arrived. Although several of the choppers had been hit, they continued to ferry in more reinforcements, extracting the wounded in return. Trooper Tobin and the Omani gunner, both seriously wounded, were CasEvaced on the first available flight. Takavesi, who suffered wounds from which most men would have died, walked calmly to the chopper without assistance. Three young Adoo prisoners, who had been captured and held in the BATT house, were also sent back for interrogation. Meanwhile, the relieving force commander, Alastair Morrison, reorganized Mirbat's defenses and with the aid of two Land Rovers, started to collect the dead and wounded Adoo. The final count of dead Adoo around the old fort was thirty-eight, but many more had been taken back by their retreating comrades.

The Dhofar War was never the same after the Battle of Mirbat. The Adoo had given it their best shot and failed, but only just. The chances that a second SAS Squadron would be just 30 miles away at the time of attack was something the Adoo could not have foreseen. Additionally, they underestimated the expertise and nerve of the SOAF pilots.

Eventually the war began to run down and the SAS forces found themselves in fairly secure positions on the Jebel from which they would send out fighting patrols to mop up the remaining enemy strongholds. They also set to work on a task at which the SAS are past masters, "hearts and minds." Small aid stations were set up and manned by SAS medics; SAS linguists would regularly talk to village leaders and their problems became the SAS problems. Civil aid teams (CAT) soon moved into the liberated areas and began drilling for water. In a land where water has priority next to life itself, the expressions on the faces of the local inhabitants were one of pure wonder and delight when the cool clear liquid gushed from the ground. Women would scream with delight, and old men would dance at the sight.

SUMMARY

The Oman War saw the SAS work together as a full squadron on several major operations. However, for the most part, they worked at Troop level. Tactics were fairly straightforward: advance to contact, fight, and hold your position. Once a position had been taken, it would be turned into a staging post for patrols so that the area could be secured. Occasionally, the squadron would mount a full-scale ambush or strong patrol behind the enemy in the hopes of catching them unaware. Several times the SAS were requested to join in much larger battles which involved throwing the full weight of the Sultan's armed forces against the Adoo.

Camp routine at "White City" during the monsoon—both rats and wild cats were a constant problem as was trying to stay dry.

CAMP ROUTINE

Once again, camp routine had several separate practices: one when in a fixed location and one when patrolling. During the time the SAS occupied the villages, they lived in houses along with the local population. Once they were up on the Jebel, they were in a fixed location which offered good defensive protection or one that had previously been built by the Squadron before. In both cases, they were all very much the same, with the SAS building a communal bunker made from sandbags, and dotted around on the high ground were individual gun-pits for defensive protection. Then larger, more permanent camps had a wire fence which was backed up by Claymore mines. Most retired soldiers will recognize this configuration as it is fairly standard.

Daily life was mostly about improving your surroundings and doing personal everyday jobs. These included building a shower, constructing a kitchen, and digging a toilet pit. From time to time there would be standoff attacks by the Adoo, but most were fairly ineffective. The majority of the time was spent patrolling.

DRESS AND EQUIPMENT

Dress in Oman was basic desert pattern—sand-colored shirt and shorts. Hats were not obligatory but many wore some type of headgear to keep the sun off. Desert boots were favored unless one planned to move over rocky country; then it was advisable to wear more substantial footwear.

Webbing (combat equipment) consisted of the standard SAS belt order with the emphasis on water, at least two bottles was advisable (at the time the camelback hydration system had not been introduced) and at least six to ten magazines of twenty rounds depending on the weapon carried. On long patrols, bergens would be carried or ferried in each night by chopper. This would contain extra ammo and water, plus food sufficient for the duration of the operation. Owing to the rocky ground, sleeping mats were essential, but few bothered to take a sleeping bag and relied on a spare blanket for the odd moments when it got cold. And it can get very cold, especially between 3 a.m. and 8 a.m. early morning.

WEAPONS

The standard weapon in Oman was the AR15, which at the time was not very reliable unless you used Remington ammo. The FN was also popular but a heavy weapon to carry in the heat; snipers carried the old but very accurate Lee-Enfield 303. Other weapons included the new LAWS 66mm disposable rocket launcher, which proved to be a great asset as did the 49mm grenade launcher.

TACTICS

At this stage of the war strategies were very basic. In the first weeks of battle, contacts would be very close and in large numbers. Rarely an hour would go by without one of the three battle groups coming under fire. It was a matter of advance to contact, hold the firefight, bring up heavy gun groups, and call in jets.

For those who have never been in a close-quarter fire-fight, it's hard to explain what it's like. Contact is generally opened by the first two opposing individuals sighting each other and the first shots fired. This then builds rapidly as all guns on both sides come into play; heavy machine guns then join in with a steady chatter, swiftly followed by the thump of mortar bombs leaving the tube. Somewhere in the middle of this one finds himself fighting in a vacuum of silence, while men shout orders and the wounded scream.

You are beyond being scared at this stage and you fight simply because you are there.

Winning the firefight is the basis of all victories. Hit the enemy with a wall of accurate fire and they will stop and eventually retreat. For the SAS, this firepower was sup-ported by two of the best infantry weapons ever produced: the 81mm mortar and the general purpose machine gun (GPMG).

Tactics involved recruiting local militia and training them to be soldiers; after which the SAS would lead them into battle. In truth it was the Firqat that were always in the lead, as they knew the terrain and were wise to the enemy tactics.

In the early days it was a simple matter of advance to contact—you got up in the morning, had breakfast, and then started forward with normally around ten Firquats and an SAS troop in the lead. Contact with the enemy would normally start in the morning and it was a matter of winning the firefight and pushing the enemy back before advancing again.

Once the majority of the Jebel had been cleared, firm bases were set up from which patrols would move into the more secluded areas deep within the mountain range. Most soldiers will recognize this pattern of events which has changed little and was used more recently in Afghanistan.

SAS SKILLS

The main skills required in Oman were that of mortar fire controller (MFC) and forward air controller (FAC). Medics also had their work cut out for them, as there were a lot of casualties, and demolitionists had to deal with booby traps or lay and remove mines. The Oman War was good for the SAS, as many of its young soldiers who had not previously been in combat gained valuable experience. It was a time when the new guys had a chance to show what they were made of and show true leadership potential. However, the main skill required by everyone was to know when to lay down suppressive firepower once the enemy had been engaged and to hold your ground before pushing forward.

HEARTS & MINDS

Once a base was established, the SAS medics and linguists would start patrolling the local villages. Any medical aid was gratefully received and helped win over the local population. Additionally, the movement into the area by local Civil Aid teams (CAT), who would produce water by drilling, help to establish long-term relations and build small villages into modern towns. CAT teams also provided agriculture and veterinary skills to improve the living standards and promote commerce.

Chapter 5

FALKLANDS WAR

While the terrain and operation could not have been different, the Falklands War was in many ways similar to that of Oman in the sense that the SAS fought alongside the regular British Army. They did not march into battle like the Welsh Guards or the Royal Marines; they fought the Argentineans in their own way as Special Forces. Much of the early work consisted of setting up observation posts and gathering intelligence on the Argentinean forces together with their dispositions. Once the war was underway, the SAS really came into their own by organizing hard-hitting raids.

The Falklands are a group of islands in the South Atlantic, which have been under British sovereignty since 1833. When the Argentineans invaded the Falkland Islands on April 2, 1982, the British Prime Minister, Margaret Thatcher, announced that Britain would win the islands back, taking them by force if necessary. Under Operation "Corporate," a task force was immediately put together and sent on its way south. Both Brigadier Peter de la Billiere and Lieutenant-Colonel Mike Rose, the Commander of 22 SAS, fought hard to have the Regiment included in the task force. By early April, members of D and G Squadrons were on their way and arrived in the Falklands three weeks ahead of the regular army.

The first objective was to find out the strengths and weakness of the Argentinean forces on the Falklands. To aid in this, G Squadron swiftly put in patrols from where they could observe the enemy. This was not the most pleasant of tasks, as the Falklands are a barren, windswept place with a good deal of rain. Owing to the swiftness of their insertion, many of the patrol were ill-equipped for long-term observation and the going was tough for the first few weeks. By day they would observe the enemy and at night they would leave their hide under cover of darkness and move several miles before transmitting their intelligence back to SAS HQ in the United Kingdom. The move away from the hide was to defeat Argentinean direction-finding equipment.

The Falklands War was fought over hostile terrain. This image is of the SAS being rescued off a glacier—a rescue that cost two helicopters. The third helicopter bravely lifted out both SAS and aircrews by a brilliant pilot.

The SAS saw their first action on the little island of South Georgia, 900 miles east of the main Falkland Isles. There was only a small Argentinean garrison and retaking the island was seen as a low-risk, high propaganda measure— with men from the SAS, SBS, and M Company of 42 Commando achieving the task on April 26, capturing the garrison and an Argentinean submarine in the process.

The SAS also mounted a diversionary raid the night before the main task force landing at San Carlos. This involved sixty men of D Squadron hitting the garrison at Goose Green with the aim of simulating a battalion-sized attack. Using a vast amount of 66mm rockets and automatic fire, the SAS laid down a ferocious barrage. Early next morning, with the main landing completed, the SAS withdrew from Goose Green.

The raid on Pebble Island was reminiscent of the attacks on the German airfields during WWII.

The early hours of May 14 saw one of the most daring raids carried out by the SAS during the conflict. A reconnaissance group from D Squadron had already reported the presence of eleven Argentinean aircraft at the Pebble Island airstrip on West Falkland. Twenty members of Mountain Troop, D Squadron, led by Captain John Hamilton, assaulted the airstrip with orders to destroy all aircraft. Despite a few last minute hitches, by the time they finished, all eleven aircraft had been destroyed or rendered irreparable; additionally several Argentinean pilots and ground crew lay dead. Two of the Squadron were wounded by shrapnel when a mine exploded, but not seriously. The success of the mission was a serious blow to the Argentine morale.

Captain Hamilton was troop commander of D Squadron Mountain Troop, and as such had been in the Falklands from the very start. One of the troop's first tasks had been to set-up observation posts on the island of South Georgia, with the goal of recapturing it from the Argentineans. The first insertion of SAS men fell to Hamilton's troop, when on April 22 they were dropped by helicopter onto Fortuna Glacier. As many an SAS man will tell you, looking at a feature on the map is not the same as the feature on the ground. Almost immediately, the Mountain Troop members realized that to say where they had been dropped would mean certain death. The katabatic winds on this glacier can easily sweep a man off his feet. The weather and exposure were so bad that they would not have lasted twenty-four hours, added to which, it was impossible to see more than a few meters in front of them. The next morning, an immediate recall was requested. The extraction cost two Wessex helicopters—both of which crashed because of the weather conditions. In what can only be described as brilliant flying, a third chopper, overladen and in blind conditions, managed to extract every man.

After this, Captain Hamilton took part in the raid on Pebble Island, as mentioned earlier. Towards the end of the war, on June 10, a patrol lead by Captain Hamilton was spotted close to Port Howard on West Falklands. The four SAS men desperately tried to fight their way out, and as two withdrew, Hamilton and his signaler gave covering fire. During this time Hamilton was hit, but continued to provide supporting fire. Some time later the overwhelming force of Argentineans killed Captain Hamilton and captured his signaler; but he had won sufficient time for the rest of his patrol to escape.

6 Troop B Squadron jumping into the waters off the Falklands prior to being infiltrated into Chile.

Not all SAS missions go according to plan, and just once in a while, mainly due to circumstances beyond their control, the operation turns into a disastrous failure.

Argentina's Super Etendarde fighter planes and their deadly load of Exocet missiles were a real threat to the British fleet during the Falklands War. They had already destroyed the HMS *Sheffield,* and the possibility of destroying a carrier was unthinkable. The only way to stop this was to destroy the Excocets on the Argentinean airbase of Rio Grande—a job for the SAS.

It was known that the airbase had five missiles, two of which had already been fired when the HMS *Sheffield* was sunk. The first plan was simple: send in a small team to infiltrate the airbase and destroy the remaining Excocets. However, as it turned out, the plan—which was hatched at Hereford—was all done in a rush with very little real planning. This was surprising, as the SAS HQ has one of the best planning teams in the world.

Under the code name Op Plum Duff, 6 Troop of B Squadron were briefed on their mission and departed at 5 a.m. on May 15, 1982. The problem was that no real thought had gone into the operation and with maps so old as to be useless, a handful of SAS were to infiltrate, via Chile, an airbase they knew nothing about. No intel on the security, perimeter fence, dogs, mines, manpower guarding the airbase, no detailed maps—nothing.

They flew first to Ascension Island in the South Atlantic, from where they were to be taken in a C130 RAF Hercules transport plane to the waters off the Falklands. Once close to the British Fleet, they dropped by parachute into the ice-cold water, only to be fished out by the navy and taken aboard one of the carriers. Once they were dry enough, they were taken by Sea King helicopter to a LZ close to the Argentinean boarder.

The first LZ was compromised and as such, the helicopter was forced to land further away from the target; eventually leaving the SAS men in the godforsaken tundra of Tierra del Fuego some 66 miles west of the Rio Grande airbase. The low-lying, gentle hills covered with vast patches of marsh and grass made it hard to move swiftly. Progress was slow, averaging less than two miles an hour. There was snow on the ground and near-freezing rain blew horizontally onto their backs. The troop commander tried to contact

Hereford and explain the situation, only to be told to get on with it.

Throughout daylight on May 19 and May 20, the men lay in their sleet-covered shelters on an undulating plain of pampas grass, covered with snow or ice. As darkness fell on the twentieth, they resumed their increasingly pointless trudge. With only two days' rations left, they were still no closer than 10 miles from the border, and from there the target was a further 30 miles across enemy territory.

By the evening of the twentieth, the Troop Commander told Hereford that a resupply was necessary and imperative before they entered Argentina. The reply was speedy and unexpected. The patrol was to head back to an emergency rendezvous manned by Captain Pete Hogg of the SAS. A meeting place was agreed—a bridge—chosen from a map with neither contours nor grid. Hogg would meet them the following night when the rendezvous would be open for only one hour after sunset.

Following that final conversation, the patrol's communications system died, finally and irreversibly.

By late afternoon on May 22, 6 Troop believed they were in the correct location. Wriggled down into the sodden undergrowth, they waited for dusk. Nothing happened, and they saw no one for the next three days. Knowing there had been a mistake, the troop Commander and one of the troopers set off for the nearest town of Porvenir, some 50 miles away. Luckily they managed to find a road and hitch a lift. Once in the town they found a small wooden hut that served as the local phone booth, from where the Commander immediately called the British Consul. The response from the Foreign office, who knew nothing about the Operation, was to give themselves up! Some choice words were exchanged and the Commander hung up.

Then, while they were considering their options, they happened to pass a local eating house; and sitting inside was Caption Hogg and two other SAS soldiers. It was not a happy meeting as they had made no effort to make the arranged RV. Disappointed, the Troop boarded a light aircraft from Santiago, finally reaching home on June 8.

The Troop Commander was disciplined for his leadership and because the operation had been a failure; he left the SAS and the Army.

On May 19, 1982, at 9:30 in the evening, elements of 22 SAS were being ferried by Sea King helicopter from the HMS *Hermes* to the assault ship HMS *Intrepid*. The flight, which covered half a mile and took just five minutes, ended in one of the worst disasters the SAS has ever known. Upon arrival at the HMS *Intrepid*, the helicopter was unable to land because another helicopter, which had landed a few moments before, needed to have its rotor blades stowed prior to being taken below deck. The Sea King made a second circuit while it waited. When it was midway between the two ships and at a height of about 120 meters, something hit the rotor blades causing it to plummet into the sea with the loss of twenty-two lives, most of which were SAS soldiers primarily from G Squadron.

For the next few weeks the SAS continued their aggressive actions, and at the end of May, D Squadron seized Mount Kent, some 40 miles behind enemy lines, and held it until relieved by forty-two Commando Marines.

The last major SAS raid was mounted in East Falkland on the night of June 14. This involved attacking the Argentinean rear while 2 Para assaulted Wireless Ridge, just a few miles west of Port Stanley. A total of sixty men from D and G Squadrons and six men from the SBS used rigid raiders to assault Port Stanley harbor, setting fire to the oil-storage tanks while laying down suppressive fire.

However, the Argentineans were fully prepared for a final assault by the British, and as the four raiding craft appeared within range, they opened fire with every weapon they had; including a triple-barreled 20mm anti-aircraft cannon depressed to their lowest trajectory. This barrage of firepower forced the raiders to immediately withdraw so as to avoid any loss of life. Despite their prompt action, all four raiders were badly hit and one even sunk, a near impossibility given their competent design. Several of the men received injuries, although none serious, and the party returned safely. The Regiment later agreed that the raid was perhaps less wise than it was adventurous.

Towards the end of the war, surrender was effectively instigated by 22 SAS's Commanding Officer, Mike Rose. Rose used psychological warfare on the Argentines, broadcasting morale-lowering propaganda with the aid of a Spanish-speaking interpreter. It seemed to have an effect because on June 14, Rose flew into Port Stanley for talks with the Argentine Generals who now realized that their cause was lost. The Argentines surrendered soon after and Rose was there to accept it, along with Brigadier Peter de la Billiere. Rumors abound that Rose did, in fact, offer the Commander of the Argentine troops, Major-General Mario Menendez, a demonstration of the firepower possessed by the Harrier jets as the Argentine troops gathered around Stanley. He is also said to have "confiscated" a valuable statue of a horse given to Menendez by the Argentine government as a token of thanks for his capture of the islands.

After the Falklands, Rose's career went from strength to strength as he became a Brigade Commander in Northern Ireland, Director of Infantry at Warminster, and Director Special Forces. He came to the forefront of public notice when he was appointed the Commander of UN forces in Bosnia on January 24, 1994.

SUMMARY

The Falkland War came as a surprise to the British, but as Prime Minister Thatcher had predicted, in the event of war we could convert much of our civilian shipping and air transport for military use. The rapid assembly of an armada soon took place and the fleet sailed south—ahead of them went the SAS.

TACTICS

The tactics throughout the Falkland War varied depending on the operation to be carried out. This ranged from setting up observation posts to raids deep behind the enemy lines. Patrol size would also change dramatically from a traditional four-man patrol to a whole squadron raid. The war was not just confined to the main population areas but ventured out almost a thousand miles to the east and with incursions into South America. The tactics for these operations are clearly defined in the true stories below.

DRESS & EQUIPMENT

Dress for the Falkland War was normal military camouflage combat dress. However, G Squadron was also the Arctic warfare Squadron and so had slightly better equipment. This included winter undergarments, good boots and gaiters, as well as fur-lined headwear that came down and covered the ears. When operating in snow conditions, white, outer winter warfare suits were worn.

Belt order equipment differed little from that of any other operational area, save for the reliance on odd equipment that would help keep you dry and warm.

Weapons

Weapons carried included the AR16 with 203 grenade launcher. Some still carried the old L42A1 sniper rifle as they had done in Oman. Other weapons included the L1A1 66mm disposable rocket launcher and L2A2 hand grenade. When required, the SAS also used the steadfast GPMG and the 81mm mortar. The SAS also acquired six American-made Stinger missiles; unfortunately the soldier trained to use the missiles was killed as a result of a bird strike on the helicopter while cross decking from ship to ship. Despite this tragedy, luck was with the SAS and the first missile fired managed to bring down an Argentinean Pucara aircraft on May 21.

Chapter 6

HOSTAGE
RESCUE

Following the hijackings at the Munich Olympics, world governments became determined to combat a new breed of terrorism. At the G7 talks the following year, the heads of government agreed to establish dedicated forces capable of dealing with any terrorist situation. In Britain, the SAS were tasked with equipping and training the new force, known by a variety of names including the anti-terrorist team, SP Team (Special Patrol), and the Pagoda Team. New equipment was purchased and the SAS responded to the training with unmatched keenness.

Today the SAS anti-terrorist team is housed in a purpose-built building which is manned by two teams, Red and Blue, twenty-four hours a day. Each team member is issued with an alert device and his movements are restricted to allow for a quick response. When a callout is initiated, the vehicles are loaded with a vast array of weapons and equipment before a briefing is given on the operational requirement. The anti-terrorist team will only move into position when requested by the home office. There are, at present, two teams, each capable of working independently or together, as the task dictates. The SAS anti-terrorist team is regarded by most of its peers to be the best in the world.

Author's Note: Captain Andrew Massey joined the SAS as a Troop captain in 1971, and was the officer responsible for devising the outline concept of the British anti-terrorist team. He was an excellent officer who went on to become a Brigadier serving as Deputy Director Special Forces.

A full SAS Counter Terrorist team circa 1985.

ASSAULT TEAM

Assault teams focus mainly on assault entry, concentrating primarily on all methods of getting into an aircraft, train, or building and closing with the enemy. They work in pairs, so that if one man is shot the other will immediately replace him. All members of the SAS anti-terrorist team spend hundreds of hours in the now famous, "Killing House," where drills and shooting techniques are honed to a fine edge.

All assault team members wear a black, one-piece fire-retardant suit, on top of which goes the body armor and weaponry. This is normally a Heckler & Kock MP5 sub-machine gun, whose clips flush across the chest. Additionally, a low-slung Browning Hi-Power pistol is strapped to the leg for backup or for use in confined spaces. In the past few years the SAS has evaluated many hand weapons, with the current model in use being the Sig Sauer P226. Respirators are normally carried in their container strapped to the back, but more likely during the actual assault the pack is discarded and the respirator is shoved up the left arm and kept for immediate use. Most actions now involve wearing the respirator: it not only protects against gas, but also presents an evil head of obscurity to the terrorists. Boots are nonslip, similar to professional climbing boots.

The assault teams have unlimited access to aircraft, buildings, trains, ships, and oil rigs during their training. Scenarios of terrorist-type sieges and hijacks are constantly played out in exercises where police and government officials take an active part.

SAS Sniper team practicing shooting on command. Respirators are worn just to make things difficult, circa 1976.

SNIPER TEAM

The SAS anti-terrorist unit has two sniper teams which can work jointly or independently, depending on the size of the problem. They work in pairs and are normally the first

to be deployed during any terrorist incident. From well-camouflaged positions the snipers will report back all terrorist movement, remaining prepared at all times to deal with any long-range situation that may present itself. Just prior to any attack by the assault teams, the snipers will indicate any terrorist targets that may be exposed. They do this by pressing a small button on the side of their weapon which in turn triggers a small light on the team commander's display board.

During a rural terrorist situation, snipers dress in a "Gilly" suit. This suit provides a foliage-type camouflage, but they will frequently dress in black—identical to the assault teams—when selecting a sniper position in a building. At times the snipers cross-train for use in the assault role.

Each sniper will have at least two rifles, one for daytime use and one fitted with a nightscope. The main sniper weapons used are the British Accuracy International PM sniper rifles, but over the years they have used various other weapons.

The snipers and command are aided by a Command & Control System which is designed to permit the unit commander to exercise split-second control over the members of his sniper team and to trigger an Immediate Action should they report hostile intentions by the terrorists (i.e., they start shooting hostages). The system operates via a highly secure digital telemetry radio link and lets the commander know when targets have been acquired. The snipers play a joint role, and depending on the situation, they can also be deployed as Assault Team members.

SUPPORT TEAM

The support team provides all the backup equipment needed, such as ladders, heavy entry equipment, and just about everything needed to set up a long-term control room. This equipment is stored in a large van which makes its way to the incident in the wake of the main team. After many years of real scenarios and endless exercises, the entire system runs like clockwork. Generally there will be some advanced warning from those first responders; this is normally the police.

The team will be put on alert and all vehicles loaded and ready to go at a moment's notice. The first to leave will be the current team commander who will fly by helicopter to the scene taking images and gathering as much information as possible. Next, he will find a location close to the incident in order to set up a holding area where the team can deploy and make ready for any immediate action. As time progresses, the support vehicle will arrive and a full operational status will bloom into life.

A rare picture of SAS Counter-Terrorist team practicing with a crossbow designed for silent killing. It was a deadly weapon, but never used on operations.

Author's Note: It is worth mentioning at this stage that within the United Kingdom the SAS Counter-Terrorist team will not move until requested by the Chief Constable of the incident area. However, all Chief Constables are invited to Hereford on an annual basis for a briefing so that they are aware of what is required in order to use the SAS should it be deemed necessary. In the United Kingdom, the police have primacy and are required to request SAS assistance. During any major incident this is normally a joint decision made by the COBRA office of the Government.

DRESS

The Assault Suit is designed to provide maximum protection against injury from heat and flame when worn with the assault undersuit. A one-piece garment worn under the Assault Body Armor, the suit is manufactured in Arvex SNX 574 flame-resistant, antistatic, liquid-repellent, black 210 gsm fabric. The areas of the forearms, knees, and shins are reinforced with quilted Arvex fabric containing Panotex flame-resistant felt, which provides additional protection against heat conduction should the wearer come into contact with extremely hot surfaces. Fitted with a drag-handle, the harness permits the wearer, when unconscious or incapable of movement, to be dragged out of the line of fire to safety.

SAS Assault team member (circa 2001) smashing through a window. A brilliant painting of the same hangs on my office wall.

The assault belt rig is designed for the SAS anti-terrorist team and carries members' personal weapons and ammunition. Manufactured in top quality black, bridle leather, it comprises a heavy-duty fully lined belt, a pistol holster, two magazine carriers, a grenade carrier holding two stun grenades, and a three-magazine carrier. Canvas versions of the same style rig are also used.

The body armor is specifically designed for the assault role, providing a very high level of ballistic protection. It is available in differing grades of ballistic protection, according to operational requirements. Front and rear ceramic plates can be inserted for extra protection if required. When the team is deployed over water there is the added protection of emergency buoyancy fitted to the body armor.

The radio communications harness is designed to permit team members to be in full radio communication with the rest of the team. An Electronic Ear Defender/Radio Headset unit protects the wearer's ears against damage from high levels of noise caused by gunfire and explosions at close proximity in confined spaces. All elements of the system are normally worn on the front of the body armor which contains a press-to-talk switch activating the body-worn microphone. Additionally, a wrist-worn press-to-talk switch is connected by a lead worn on the wrist or on the hand inside the glove, enabling the wearer to press it against his weapon and thus activate the body-worn microphone.

The Voice Projection Unit is a miniature amplifier unit designed for use with S-10, SF-10, and CT-12 Respirators. Powered by a standard 9-volt PP3 type battery, the unit is rapidly and easily attached to the diaphragm on the respirator face-piece. When activated by a switch, the unit projects the wearer's voice over a distance of some 30 meters, thus enabling hostages or other personnel in the target area to hear clearly any commands being given during an assault. It sounds a bit like Darth Vader speaking.

RANGE ROVER

Range Rovers were first used by the SAS when the anti-terrorist team was formed in November 1972. The government sanctioned the purchase of six Range Rovers, and a team of SAS soldiers was sent to the factory to collect them directly from the assembly line. The characteristics of the Range Rover, which at the time was only a year old, were ideally suited to the role; the drop-down tailgate allowed for easy loading, and the vehicle itself could be used in an Immediate Action (IA). Twenty-five years on and the Range Rover is still used by the SAS, having been adopted as a main assault delivery vehicle. Platforms and ladders attached to the Range Rovers can carry the assault personnel directly to the required height of an aircraft door or building window.

Entry ladders came in all sizes, each one to fit the job it was designed for. The double ladders allow one member to open the door while the other can make a rapid entry.

ASSAULT EQUIPMENT

The range of assault equipment used by the SAS is vast. Most items have been designed to fulfill a specific task: either protection or speed entry. Products range from body armor and stun grenades to special cartridges that can rip off a door hinge without causing damage to anyone inside the room. The vehicles are equipped with the basics needed for an Immediate Assault while the rest of the equipment will be delivered by the support team.

STUN GRENADE

Royal Enfield Ordnance, at the request of the SAS, had experimented with various devices, and the one that seemed most favorable was the stun grenade. The device is a G60 which produces a loud noise (160 Db) combined with high light output (300,000 cd) without any harmful fragmentation. This is capable of stunning anyone in close proximity for a period of around three to five seconds when detonated and is one of the most effective items in the anti-terrorist armory. The effect is not dissimilar to the flashing strobes in a disco, but a million times more effective.

Also called a "flash-bang," this effective but nonlethal device has become essential in almost all hostage-rescue scenarios. Originally developed for the Regiment, it has now become a standard item in many of the world's counter-terrorism squads. It contains a mixture of magnesium powder and fulminate of mercury, which detonates once the ring is pulled and the grenade thrown. The grenade bodies are constructed with a minimum of metal parts to ensure that there is no danger of hostages being injured by fragmentation.

Grenades were used for the first time, operationally, at the Mogadishu hijack, where three were thrown just prior to the assault. During the Iranian Embassy Siege it is suspected that one of the grenades set fire to a curtain—although this was never proved—resulting in the building being gutted by fire.

THERMAL LANCE

The Thermal Lance is designed for cutting mild steel, including objects which are under water. The basic system consists of a 12-foot flexible, thermal lance made from Kerie cable, a single three-liter oxygen cylinder fitted with pressure gauges, a pressure regulator, battery-powered igniter, and a three-way valve which switches the system's working pressure on or off. Once ignited, the Kerie cable burns at approximately two feet a minute during cutting, which gives a maximum cutting time of six minutes. The SAS anti-terrorist team carries a backpack-portable system that weighs 10.5 kilograms (about 23 pounds), which is used for cutting during assault entry.

RAPID ENTRY EQUIPMENT

There are two types of entry: covert and overt, plus something in between. The SAS have a wide range of equipment designed to enable rapid entry through doors, windows, and walls of buildings. Included in this range are silent hydraulic cutters and spreaders as well as a range of rams, crowbars, and axes. These include the following:

- Manual Door Ram
  A handheld ram designed to force open inward-opening doors by being swung against the lock area and imparting a weight load of approximately three tons. It is effective against all but reinforced steel doors. Weight: sixteen kilograms (about thirty-five pounds).
- Door Ripper
  A lightweight tool designed to force outward-opening doors with the blade being driven between door and frame in the area of the lock. A ratchet mechanism aids the overcoming of resistance by allowing the blade to be worked behind the door to provide increased force.
- Hydraulic Door Ram
  Designed to force reinforced inward-opening doors and supplied with three sets of claws to suit all standard widths of door from 760 to 920 millimeters. The main ram is positioned over the lock area while the secondary ram forces the jaws into the frame. Operation of a valve activates the main ram to force the door open with a maximum force of five tons. An eleven-ton version is also available.

In addition to rapid entry tools, the SAS use an extensive range of assault ladders of differing widths to cater to the majority of operational requirements during a siege situation. These include single section, multi-sectional, and extending types in single width, double width, and triple stile designs. All ladders are manufactured from structural-grade aluminum alloy with deeply serrated rung sections and heavy-duty rectangular sections. All are built to order and thus nonstandard designs and lengths are available. All ladders are fitted as standard with nonslip rubber feet, noise reducing buffers on all exposed faces, and are finished in black polyester powder coating with etch primer.

The modern "Wall Breaching Cannon" can instantly make a hole in a wall without doing damage or injury to possible hostages.

These ladders are available to the SAS in single and double widths and triple stile designs up to four meters in length. They offer silent climbing and are ideally suited for gaining rapid access to public transport vehicles, ships, aircraft, or for scaling walls. Wall hooks and sniper platforms can be fitted to all sizes.

Wall Breaching Cannon (a.k.a. Harvey Wallbanger)

The Wall Breaching Cannon is a device that eliminates the need for using high explosives as a method of entry in a hostage situation. Every wall differs and it is very difficult to judge the amount of explosive required to blow a hole without causing severe debris on the opposite side that could potentially wound or kill innocent hostages. The SAS have no second chance during any assault, thus, for walls of unknown strength, more explosive than necessary is invariably used, compounding the undesired effects. The wall breaching cannon is a more suitable alternative to using high explosive.

It is designed to direct a heavy, soft projectile with sufficient velocity to accumulate enough kinetic energy to breach a wall, then instantly dissipate the energy after breaching. A water-filled plastic container is fired by compressed air and fills this requirement very adequately. The device designed to launch the container is a muzzle-loaded smooth bore barrel. The rear of the barrel is fitted with an air reservoir separated from the main barrel which when triggered by six hundred pounds psi, will create a hole large enough for a man to step through.

IMMEDIATE ACTION

The Immediate Action (IA) as it is known is carried out in response to any unforeseen actions by terrorists. From the moment the SAS Counter-terrorism team arrives at the incident area, they will prepare for this. The Immediate Action (IA) plan will allow the team to close in on the terrorists in order to prevent any further loss of life or destruction of property. It involves getting the very basic of information, (i.e., what is the latest situation, where is the enemy, what are their numbers, how many hostages do they have). From this information, a basic assault plan will be drawn up ready to go at a moment's notice should the terrorists start doing something stupid like killing hostages.

Author's Note: As information comes into the holding area it is displayed on a monitor outlining everyone's assault positions. These positions and orders can change several times within minutes and soldiers are required to keep abreast of the rapidly changing plan. Normally, the IA will become the foundation for a more positive assault plan that has a better chance of succeeding.

The SAS Counter-Terrorism team has a fearsome reputation for not taking prisoners during its assault; in truth it is difficult to react any differently when faced with a split second decision to make between friend or foe.

On December 6, 1975, after attacking a restaurant, a four-man Active Service Unit of the Irish Republican Army took a middle-aged couple hostage in a flat in Balcombe Street, London. Armed police were on the scene almost immediately and Scotland Yard was able to monitor the proceedings in the flat using fiber optics. However, the gunmen held their ground and the siege, now in its eighth day, looked as if it was going to continue indefinitely. Inside the flat, the terrorists were able to listen to a transistor radio and heard a report from the BBC suggesting that an SAS unit was preparing to intervene and take over the building. Hearing this, the terrorists immediately surrendered to the police.

HOSTAGE-RESCUE SCENARIOS

Terrorists can attack in any number of ways; they are normally savage and media seeking. For example they may take hostages and hold them to ransom; alternatively they may simply commit murder by bomb or bullet. It really depends on the situation and what the terrorist organization is trying to achieve. If the attack is a reprisal, then it is almost certainly going to be a bombing or shooting incident. If the organization is looking to have fellow prisoners released from jail or have their grievances recognized by the world at large, then they will almost certainly result in some form of hostage blackmail. To prevent such actions from taking place, the security forces monitor most of the indigenous terrorist organizations in order to anticipate any future actions they may be planning.

Where the terrorists carry out a bomb or shooting attack the security forces can only react in retrospect. Only when prior information is obtained can counter-measures be initiated to prevent or limit the damage caused by such an incident. On the other hand, when a siege situation or hijacking takes place, the security services are in a much better position to respond. To the anti-terrorist unit, these actions can be defined as a number of options: aircraft hijacking, building assault, vehicle assault (including trains and coaches), and seaborne (including ships and oil rigs). Outline contingency plans for all these events already exist and most can be modified to a specific incident.

Prince Charles and Princess Diana going through SAS hostage training in the killing house; live rounds are fired within inches of their bodies.

ROYAL FAMILY AND SENIOR POLITICIANS

Being prepared is paramount to any successful counter-terrorist operation. Not just training the team but also training those who are main targets for terrorism. In the United Kingdom, this involves preparing members of the Royal family and senior politicians such as the Prime Minister. While the Royal family has the Royal Protection Group, the SAS maintain a role whereby they may be called to deal with a situation outside police control. In keeping with their doctrine, it is wise to let any VIP know what could be in store for them should they be kidnapped and taken hostage. Thus, most VIPs are invited to Hereford where they will participate in hostage rescue training, leaving them with a better understanding of what can be achieved.

AIRCRAFT HIJACKING

The hijacking of an aircraft presents two possibilities: the threat of killing hostages or the threat of converting the aircraft into a flying bomb. Most hijackings take place in the air which favors the terrorists' chances of success, in as much as there is very little resistance to their initial attack. This is mainly because the passengers are too scared and the crew is too busy flying the aircraft. In addition, any damage to the aircraft while in flight can produce fatal results.

The standard terrorist procedure is to smuggle guns, grenades, or explosives onboard the aircraft and then use these to take command using the threat of violence against the crew and passengers or in the extreme—total aircraft destruction. Demands at this stage are normally simple: "recognize that we are in charge and take us to wherever we want to go."

Most aircraft carry just the right amount of fuel for the required journey, which under normal circumstances means they must land either before their planned destination or at it. Once on the ground the situation changes, as it allows the anti-terrorist team an opportunity to assault the aircraft. If time permits, contingency plans are put into operation, and diverting other aircraft in the interest of safety stops the terrorists from changing aircraft. Additionally, snipers can be deployed to prearranged hides prior to the aircraft landing and assault teams can prepare for an IA should the situation become unstable.

Any assault is divided into three parts: the approach, the entry, and the assault. With few exceptions, an assaulting team can approach most aircraft from their blind area at the rear. A good team commander will know exactly what area he has to maneuver in. This silent approach allows the team to reach the outside of the fuselage undetected. Once in position, ladders are placed against the doors and entry points prior to an entry being made.

The size of the aircraft involved and the number of passengers onboard will dictate how many entry points the team needs to breach in order to make a successful entry. Almost all aircraft doors, both normal passenger and emergency, can be opened from the outside of the aircraft while it is on the ground. This is a design element which facilitates emergency services access in the event of accident, crash or other emergency.

Aircraft hijacking requires a knowledge of where to gain entry from the outside of an aircraft.

Most aircraft, especially the larger types, require that the assaulting team be elevated several meters in order to gain access. A wide variety of ladder systems are available to the assaulting team, and these can be adjusted to reach any type of aircraft door. In some cases, two ladders are used side by side and placed against the aircraft fuselage. The idea is that the first man operates the handle and uses his body weight to swing the door open. This technique normally requires two hands and blocks the view of that particular team member; however the second member is free to fire or enter the moment the door is ajar. Ladders are also used to gain access to the wings, whereupon the emergency doors can be opened.

Until recently, stealth has been the most popular method of assaulting hijacked aircraft. However, many anti-terrorist teams now favor a rapid-response vehicle. This vehicle has a preassembled platform system attached which can be adjusted to any height. The result is a modern mobile siege tower which transports the assaulting members at the correct elevation directly to the aircraft entry points. While the vehicles can still take advantage of the blind area at the rear of the aircraft, they also offer a rapid delivery. It is incumbent on the team commander to get all his men onto the aircraft in order to facilitate a good assault; therefore the vehicle is normally armored.

It is not possible to give details of all types of aircraft, as there are so many sizes and variations. However, my personal plan to assault flight LH181 while on the ground in Dubai is a good example.

The Boeing 737 is a simple little animal where anti-terrorist drills are concerned. There are only three options for entry: tail, wing, and front catering area. I thought that if the terrorists began to carry out any threatened shootings, they would naturally take the precaution of covering the main doors. It seemed less likely that they would cover the two emergency exit doors leading to the wings, so the plan that basically fell into shape was to attack through these. The fact that the wing emergency exits were designed to be easily opened from the outside was another strong factor for us adopting this mode of attack—and there were others. The German anti-terrorist team, the GSG9 (Grenzschutzgruppe 9), had also discovered a blind spot where the wing joins the aircraft body. Two men could sit beneath the emergency doors and not be visible from any of the windows.

By comparison, the entry and exit points at the front and rear require considerable manhandling and some time to get them open. For instance, the front door is operable through a small hatch on the outside of the plane which allows the door to be opened and brought down and the stairs to extend automatically. The basic moves involved in my plan were:

1. To make a single-file approach to the aircraft from its blind spot at the rear, assemble our ladders quietly, and erect them to the wings and the rear door.
2. To put each of the two leading assault teams covertly on the wings—one outside the port emergency exit and the other outside the starboard—with the second assault pairs waiting on the top rungs of the ladders. Each of these assault teams consisted of an SAS man or a GSG9 man backed up by one of the best of the soldiers from the Palace Guard. The reason that we involved the Palace Guard in this way was purely political. (The Dubai Government being unwilling to allow us to mount the assault unless they played a part.)
3. To position backup squads beneath the rear area of the plane ready to scale the ladder, open the rear door, and effect entry as quickly as possible. At the same time, a second backup squad would move quietly forward until they were beneath the front door area, ready to erect their ladders and follow suit. The backups would coordinate their moves with those of the assault teams.
4. When everyone was in position and the "GO" given, the leading assault teams were to stand, punch the emergency exit panels, and drop the doors onto the laps of the passengers in the mid-section of the cabin. The teams would then enter the port side pair clearing to the front of the cabin, the starboard team clearing to the rear. The leading teams were to receive immediate backup from the second assault pairs entering behind

them from their stations at the tops of the ladders to maintain control of the center of the aircraft.

5. Simultaneous with the assault, the outside squads were to open both front and rear doors and enter the plane. The intention here was to provide further backup in case of any problems and also provide routes for the hostages to leave the plane, which by this time would be full of smoke from the stun grenades.

As I mentioned earlier, the 737 is a fairly simple animal. Once entry has been effected to the center of the aircraft, the starboard assault team gains a clear line of sight to the restroom doors at the rear of the cabin. The port team, moving forward through the economy area, arrives in the first class section which leads into the front catering area. Directly beyond this is the flight deck—the door to which is usually closed. The only obstacles the team would encounter are this door and the curtains that separate economy and first class. Although this basic plan is quite uncomplicated, we calculated that it would require a great deal of practice to get the timing right—particularly the time it would take the assault teams to effect their entry and make their way to the front and rear of the passenger cabin. I anticipated that as soon as we dominated these points, the only people in serious danger would be the crew in the cockpit.

Iranian Embassy with SAS troops on the front balcony about to blow the door in.

This plan worked well during rehearsals in Dubai and was adapted for the final assault in Mogadishu—which was successful.

BUILDINGS

Assaulting a building depends of several factors: how many floors need to be cleared, the location and number of both terrorists and hostages, main entry points. In most cases it is possible to get close to the main entry points covertly, and opening the access points can be done quietly or with a great deal of noise depending on how quick the team needs to get in. A perfect example of this was the Iranian Embassy siege in London

At 11:25 a.m. on the morning of Wednesday, April 30, 1980, six armed gunmen took over No. 16 Princess Gate, the Iranian Embassy in London's Kensington district. It was learned that the terrorists were opposed to the regime of Ayatollah Khomeini and were seeking the liberation of Khuzestan from Iran. As they took control of the embassy, they gained twenty-six hostages including a British police-man who had been on duty at the entrance. This was some-thing that might have gone unnoticed, but for the fact that minutes later a burst of machine-gun fire could be heard. The police were on the scene immediately; swiftness initi-ated by the captured policeman, Trevor Lock, who had man-aged to alert his Headquarters before being taken by the terrorists. Armed D11 marksmen soon surrounded the building and the siege negotiating plans were put into operation.

By 11:47 a.m., Dusty Grey, an ex-D Squadron SAS man who now worked with the Metropolitan Police, was talking to the Commanding Officer in Hereford. His information contained the briefest details, but it was enough to alert the regiment. Several minutes later, the duty signaler activated the "call-in" beepers carried by every member of the anti-terrorist team. Although the SAS had prior warning, there can be no move before an official sanction from the Home Secretary, who at the request of the police will contact the MoD. Despite this red tape, it makes sense for the SAS to think ahead, and positioning the anti-terrorist team closer to the scene can save time. Around midnight of the first day, most of the team had made their way to Regents Park Barracks, which had been selected as a holding area. From here various pieces of information could be assembled and assessed. A scale model of No. 16 Princess Gate was ordered to be constructed, a task which fell to two pioneers—drafted in from the nearby Guards unit. Additionally, an intelligence cell was set up to gather and collate every snippet of information that would aid an assault.

By this time the terrorist leader, Oan, had secured his twenty-six hostages and issued his demands. These included the autonomy and recognition of the Arabistan people, and the release of ninety-one Arabistani prisoners. The line taken by the terrorists was hard but fair, and despite several threats to blow up the embassy and kill the hostages by Thursday, May 1, they had released a sick woman. Later that same day, Oan had managed to get a telephone call through to Sadegh Ghotzbadeh, Iran's Foreign Minister. The conversation did not go well; Oan was accused of being an American agent and informing that the Iranian hostages held in the Embassy would consider it an honor to die for their country and the Iranian revolutionary movement.

Around this time, Chris Cramer, a BBC sound man, had become sick with acute stomach pains. His partner, BBC sound recordist Sim Harris, pleaded with Oan to call for a doctor immediately. This was done, but the police refused to comply, and in the end, Cramer was released, whereupon he stumbled out of the embassy door and into a waiting ambulance.

Later that night, again under the cover of darkness, three Avis rental vans pulled up in a small side street by Princess Gate. Men carrying traveling cases quickly made their way into No. 14, just two doors down from the embassy. Within minutes they had laid out their equipment and made ready for an Immediate Action (IA). At first this was very simple: if the terrorists started shooting, they would run to the front door of No. 16 and beat their way in—a slow and primitive action, but better than doing nothing until a clearly defined plan could be organized.

By 6 a.m. on the morning of Saturday, May 2, the situation inside No.16 was getting very agitated. Oan rang the phone which had been set up between the embassy and No. 25 Princess Gate—the Royal School of Needlework—which now housed Alpha Control (main forward control point) and the police negotiator. Oan's main criticism was that no media had been broadcast about the siege, so how could his cause be heard? By late afternoon on the same day, Oan was allowed to make a statement which was to be broadcast on the next news slot—in return, two more hostages were released, one of whom was a pregnant woman. The trouble was Oan would not release the hostages before the statement was read; likewise the police wanted the hostages first. In the end a compromise was reached, and the broadcast went out on the evening nine o'clock news.

Two hours later, eight members of the SAS team had climbed onto the rear roof of No. 14 and were making their way amid a jungle of television antennae to No. 16. Two of the men made their way directly to a glass skylight and after some time, managed to get it free. They found that it opened directly into a small bathroom on the top floor of the Iranian Embassy, and would provide an excellent entry point. Meanwhile, other members secured ropes to the several chimneys and made ready for a quick descent to the lower floors where they could smash in through the windows.

By 9 a.m. on Sunday morning, things seemed to be heading for a peaceful settlement. Oan had agreed to reduce his demands and at the same time Arab ambassadors had attended a Cobra Committee (Cabinet Office Briefing Room) meeting in Whitehall. This committee was chaired by the home secretary, William Whitelaw, who was to all intents and purposes in charge of the whole operation—deciding the path of any action. He was aware the SAS anti-terrorist team now had access into the embassy and that all efforts were being made to penetrate the wall. To aid this, various sound distractions supplied by the Gas board working in the vicinity avoided the drilling being heard. COBRA was also aware that a basic plan had been formalized. This plan involved attacking all floors simultaneously, with clearly defined areas of demarcation to avoid over shoot. Mock-ups of the floor layouts were constructed from timber and burlap sheeting and assembled at Regents Park barracks in order that the SAS could practice.

The police, who had adopted a softly negotiating approach managed to keep control of the siege for several days' time. That was desperately needed for the SAS to carry out covert reconnaissance study plans, build models, and more importantly: locate the hostages and terrorists within the embassy building. A major break was talking to the hostage Chris Cramer, the BBC sound engineer. It was a big mistake by the terrorists: in his debriefing to the SAS, he was able to give them precise and detailed information about the situation inside the embassy.

By the sixth day of the siege, May 5, the terrorists were becoming frustrated and the situation inside the embassy began to deteriorate. All morning, threats were made about executing hostages, and at 1:31 p.m., three shots were heard. At 6:50 p.m., more shots were heard, and the body of an embassy press officer was thrown out. The police immediately appeared to capitulate, stalling for time, while SAS plans to storm the embassy were advanced. At this stage the police negotiator worked hard to convince the terrorist leader not to shoot any further hostages and that a bus would be there shortly to take them to the airport, from which they could fly to the Middle East. During this telephone conversation, the SAS took up their start positions.

A handwritten note which passed control from the police to the SAS was handed over. Shortly after, while a negotiator from Alpha control talked to Oan, the SAS moved in. Oan heard the first crashes and complained that the embassy was being attacked. (This conversation was recorded, and one can clearly hear the stun grenades going off; Oan's conversation is cut short by a long burst of machine-gun fire.) For the assault team, the waiting was over, and the words guaranteed to send their adrenaline pumping were given. "Go. Go. Go."

At 7:23 a.m., eight men rappelled down to the first floor balcony, with ropes secured from the embassy roof. The assault came from three directions, with the main assault from the rear. Frame charges were quickly fitted to the windows (by now they had been perfected) and blown. Stun grenades were thrown in advance of the assaulting men and the SAS went into action. Systematically, the building was cleared from the top down, room by room. The telex room on the second floor, which housed the male hostages and three of the terrorists, was of utmost priority. Realizing that an assault was in progress, the terrorists shot and killed one hostage and wounded two others before the lead SAS team broke into the room. They immediately shot two gunmen who were visible, while the third hid among the hostages and was not discovered until later. As rooms were cleared, hostages were literally thrown from one SAS soldier to another, down the stairs and out into the back garden. At this stage they were all lain facedown on the ground while a search was conducted for the missing terrorist.

Breaking the siege took just seventeen minutes. The SAS took no casualties, other than one man who got caught up in his rope harness and was badly burnt. Once the embassy had been cleared and all the terrorists and hostages had been identified, the problem was handed back to the police. Meanwhile, the SAS vacated No. 14 and went back to Regents Park barracks in time to watch themselves on television. Still dressed in assault gear and clutching cans of Fosters lager (someone was on the ball), they crowded around, eager to see themselves in action. Halfway through, Prime Minister Margaret Thatcher, who had left a dinner date, arrived to personally thank the SAS. She circulated, as one man put it "Like a triumphant Caesar returning to the Senate," her face glowing with pride and admiration at her Imperial Guard. Then as the news started to show the full event, she sat down on the floor amid her warriors and watched with them—there can be no greater approbation from one's leader.

In total, there were twenty-six hostages taken in the Embassy when the siege started. Of these five were released before the SAS assault. Two died, but the remaining nineteen survived. Of the six terrorists, only one survived.

SHIPS AND OIL RIGS

Ships and oil rigs are assaulted in the same way, except when a ship is tied up alongside the dock—then it is treated more as a house assault. In recent years there have been a lot of ships taken at sea, mainly by pirates seeking to hold the crew and cargo to ransom. Assaulting and gaining access to a moving ship is extremely difficult, even for Special Forces, and it is amazing that more pirates are not killed in the attempt.

Both ships and rigs have the advantage of good visual range where they are able to see any advance towards their location, which for a ship in dangers of being raided by pirates is a good thing; in the event of an assault force wishing to stage an attack it is not so good.

Gaining access to a high structure in the sea is difficult and requires several steps. First the team needs to be taken as close as possible to the target without being seen. Secondly they need to climb up the ship of structure in order to perform their assault obligations. Both of these allow the terrorists to form some kind of retaliatory action of rejection.

Assaulting a ship from the sea surface is not as easy as one would expect.

The main methods of approach are either by parachute or by surface boat, both of which are highly visible. A sub-surface assault by divers is the most covert method, however getting from the sea surface to a point where the terrorists can be engaged is difficult.

Access to high levels can be reached using a special grapple and a flexible ladder which is hoisted into position. The actual assault is carried out in a similar way to assaulting buildings with each deck being cleared in turn. The cruise ship *Achille Lauro* was hijacked in October of 1985, during which an American holiday maker was murdered. The perpetrators having achieved their aim sank the ship and escaped on an Egyptian airliner.

TRAIN AND COACH ASSAULT

The SAS were once provided with a full train plus several carriages by British Rail; there was only one stipulation, which was that the wheels had to remain on the track. It was a very worthwhile exercise and many lessons were learned. The primary task for either a train or bush hijacking is to stop the vehicle so that a safe assault can take place. Trains can be diverted and put in a position where they must stop or derail. While this does point to an assault about to take place and warn the terrorists, it is better than trying to land assault troops on a moving train. Once access to the train has been made, the carriages are cleared by multiple teams, normally one per carriage.

While terrorists rarely hijack buses, they are used to transport hostages from one point to another. This is a critical point in any counter-terrorist scenario, as it offers the team a clear chance to engage with the terrorists. Whenever possible the coach is supplied by the team to fulfill any requests by the terrorists to move hostages. The team will have more than one coach available in case the first is rejected by the terrorists.

The coach will be stopped precisely at a predetermined point where the assault team will be waiting. (I cannot detail how this is done as it still remains a secret.) Once stopped, the team will deploy a distraction and place short ladders against the coach body while some make an entry. Windows are smashed using baton rounds, making it easy for the assault team members to shoot any terrorists inside. Stun grenades are deployed providing the team with the vital few seconds needed to secure the hostages.

Coach assault using a helicopter for delivery of the team. If the timing is right, this method of entry works well.

As both means of transport are fairly narrow, it is not always necessary to enter the coach or train carriage in order to pacify any terrorists. If things start to get out of hand, CS gas will provide an excellent means of distraction. The Depont Train hijacking was one of the earliest instances of just such a terrorist act.

A train traveling between Assen and Groningen in Holland was hijacked by nine Moluccan terrorists on May 23, 1977, with fifty-one hostages being taken. At the same time, 110 hostages, mostly children, were seized at an elementary school at Bovensmilde, although 106 were released a few days later with a stomach virus. The standoff lasted for three weeks until the psychiatrist who was conducting negotiations on behalf of the Dutch government disclosed his concern that the terrorists were about to start killing hostages.

The Royal Dutch Marines Close Combat Unit prepared to storm the train at 4:53 a.m., working in five-man teams. SAS advisors were in close contact throughout the hijacking and recommended the use of stun grenades, but the Dutch decided against this form of action, preferring instead to instigate the assault under the distraction of six F104 Starfighters flying low over the train. Six of the terrorists were killed and three surrendered. Two hostages, who panicked when the firing began, were killed. One other hostage was also shot, albeit not fatally.

In a simultaneous assault at the school, the results were much better. Three of the four terrorists were caught unawares, asleep in their underwear, when the Marines broke through the school wall with an armored vehicle. All four terrorists were captured and the hostages were safely released.

SUMMARY

The techniques and skills are clearly depicted in the above text save for the ones that remain in service today, and it would be wrong to explain those. One thing that I originally found odd was the way in which most of the world's counter-terrorist teams had evolved in a similar fashion. This basically started with the Europeans, who without prior knowledge of their counterparts in other countries evolved along the same lines. For example, in Germany the anti-terrorist unit was originally developed from the border police Grenzschutzgruppe 9 (GSG9)—similarly in France, their Groupement d'Intervention de la Gendarmerie Nationale (GIGN) is also a police unit. Whereas in the United States, Delta Force is formed from its special forces units, similar to that of the British SAS. Yet each team had identified comparable needs in order to create a counter-terrorist team.

Assault troops, sniper units, methods of entry, and deployment vehicles are all very similar, and once the teams had gotten to know each other, the sharing of ideas and equipment propagated. One such idea developed by the British Government was the stun grenade, an item that was quickly adopted by many others. Likewise the similarity in weaponry—the Heckler and Kock MP5—became the symbol of the modern counter-terrorist team during the 1970s.

Today there are numerous exchange programs between the various counter-terrorist teams around the world, each learning from each other.

Chapter 7

NORTHERN IRELAND

Trying to understand the troubles in Northern Ireland is not easy, as they stretch back to 1167 when the English first appeared in the area. While they were never completely reconciled with the local populous, they did at least live side by side. It was not until 1534 when a Catholic revolt was led against the English Protestant King in Northern Ireland that the troubles really began. The rebellion was swiftly put down and more land was confiscated and given to English, Scottish, and Welsh Lords; in 1650, Cromwell sent thousands of his Parliamentary soldiers to settle in Northern Ireland, hence large populations of Protestants were at odds with the indigenous Catholics.

By the start of the twentieth century, the lines had been drawn with the Protestants in the North and the Catholics in the South. It also saw the formation of various groups such as Sinn Fein in 1905 and the Ulster Volunteer Force (UVF), with large bodies of armed men on both sides. During World War I, the Easter Uprising of 1916 caught the British off-guard, but was swiftly and brutally put down; this only served to make martyrs of the executed leaders and the movement unified even more strongly.

By May 1921, the country was split with six predominantly Protestant counties in Ulster becoming known as the North and the remaining 26 counties forming the South. At the same time, the UVF vowed to continue its fight and launched attacks on the newly formed Irish Republican Army (IRA). And thus, Northern Ireland was born.

The South gained its independence and for a while there was relative peace, but mainly due to the economic decline in the 1960s, trouble started to erupt between the Catholics and Protestants in the North. It came to a head in 1966 on the fiftieth anniversary of the Battle of the Somme and the Easter Rising, with both sides spilling blood. A small force of British troops was sent to keep the two sides apart. Slowly but surely, the British Army became the main focus of attention for the Irish Nationalists.

IRA ACTIVE SERVICE UNIT

An Active Service Unit is an operational cell of the Irish Republican Army. These cells usually consist of three to four men, but this number may increase to ten in certain circumstances. Their purpose is to carry out bombings, mortar attacks, assassinations, and strikes against the security services, prestige targets, and individuals, both in Northern Ireland and on the British mainland. Each unit operates independently, thus cutting the risk of any disloyalty or double-dealing, which would affect the whole organization. Weapons and explosives will be supplied by the member designated to be quartermaster, and a set of trusted go-betweens will normally ferry such equipment around. Cell members are normally well known to the security forces and are constantly stopped and searched. If a member of the cell is killed, then the next most trusted go-between is normally moved into his place.

While the riots became ever violent, it was Bloody Sunday that caused a grievance which lasts to this day.

On January 30, 1972, the Parachute regiment suppressed a riot during a civil rights march which resulted in the deaths of fourteen protesters; the incident became known as "Bloody Sunday." Violence in Northern Ireland continued to escalate and at last the SAS became involved. The SAS entered the conflict in 1973, but most teams were already there attached to a local regiment—trying to understand the situation and to see if there was a role for them.

In 1976, the SAS were to become a political tool used by the British government to show the IRA it had teeth. Results came quickly: on March 12, 1976, Scan McKenna, a known IRA member, was lifted from his home over the border in Eire. He was dragged from his bed and frog-marched into Northern Ireland, whereupon he was handed over to the Royal Ulster Constabulary (RUC). Less than a month later, another IRA member, Peter Cleary, was lifted from the home of his fiancée, who lived just north of the border. The house had been under observation for some time as it was known that Cleary was soon to be married, and it was just a matter of waiting. Once the suspect had been taken into custody, the SAS men moved to a pickup point and waited for their helicopter. However, Cleary tried to escape, and was shot dead as he did so. The incident did not go down well. The IRA claimed that he had been murdered.

Several senior officers in Northern Ireland were aghast at having a bunch of men such as the SAS within their midst, but as the Prime Minister had personally sent them, there was little they could do. However, the situation did serve to send a clear message to the IRA: border or not, we will come and get you. This theory was confirmed when, at 6:00 p.m. on May 5, 1976, the Eire police stopped two men in a car at a checkpoint within the Republic, and in the process discovered that it had members of the SAS on its hands. To make matters worse, two backup vehicles arrived to assist the first, and in total eight SAS men, together with their vehicles and weapons, were taken into custody by the Eire police. As the news emerged, all hell broke loose and the newspapers had a field day. The men were taken to Dundalk and then on to Dublin. All were charged, and in the end it was just as embarrassing for Eire as it was for the SAS soldiers, most of whom were fined £100 each for having unlicensed weapons.

NORTHERN IRELAND CELL

This special cell was set up to develop skills for the SAS to use in Northern Ireland prior to deployment. It was little more than a wooden hut and a few covert cars but it provided valuable training. The original NI Cell was commanded by Major Tony Ball and a staff of four SAS NCOs, all of whom had been sent to the Province to gain firsthand experience from the regular British Army. Members who had not served in Northern Ireland were taught skills such as covert car surveillance, advanced observation post routines, and facial memory of IRA suspects.

The original NI Cell was located in the old seamstress building at the back on the cookhouse, in what was then known as Bradbury Lines, but was phased out once all the Squadrons had rotated through Northern Ireland and the Regiment had established its own methods of operating. Much of the training consisted of vehicle and helicopter surveillance exercises in and around the Bristol area, which was not dissimilar to that of South Armagh.

Excellent photo of the Northern Ireland Cell practicing anti-ambush drills at Camp One. Ammunition being fired here is live.

FOUR SQUARE LAUNDRY

Four Square Laundry was an operation set up by a force known as the Military Reconnaissance Force (MRF), an undercover unit that functioned in Northern Ireland in the early 1970s. One of their tasks was to operate a mobile laundry service, collecting from house to house. They were assured of good custom, as their prices were far lower than their nearest rivals. Prior to washing, all the clothes would pass though a machine that carried out a forensic test for explosives; those found would indicate where bombs were being assembled. It ran successfully for several years until a member of the MRF who had been a former member of the IRA and converted to work for the British, changed his allegiance back. This led to a Four Square van being shot up. The male driver was killed but the woman managed to escape; both were undercover British soldiers.

The work was extremely dangerous and many of its operators were killed; the first was Captain Anthony Pollen who was executed in front of a large crowd of people after being discovered working undercover at a Sinn Fein event.

ACCOMMODATION

When securing a location in Northern Ireland, the SAS would fit in as close as possible to the action. Whenever possible, the SAS would hide themselves within the same camp as a regular British Regiment stationed in the province. However, while they shared the same camp, the SAS camp area would be segregated and secured behind a wire fence so that secure access was always in place.

Within this compound the SAS would have their own accommodations, operation rooms manned 24/7, several briefing rooms, mess hall, bar, and stores. A car pool would supply all the covert cars and helicopters that were on hand from the camp general air support.

At the start of the troubles, the SAS used segregated accommodations in Bessbrook Mill. The large disused mill had been taken over by the British Army and housed the local Battalion during its tour of duty. SAS patrols would operate from here into the border countryside of Crossmaglen and Forkhill. While covert cars were sometimes used for drop-offs and pickups, RAF or Army Air Corps would airlift most patrols.

Bessbrook Mill home to the SAS when they first arrived in Northern Ireland. The building and helicopter pad came under IRA mortar fire several times.

These lone operatives often found themselves exposed by the IRA and having to make a break for it. Captain Robert Nairac got himself into such circumstances in Northern Ireland in May 1977. When working alone he was discovered, captured, and murdered by the IRA. Having been educated at Ampleforth, the top Catholic public school, and

Lincoln College, Oxford, Nairac joined the Grenadier Guards. He was both intelligent and a keen sportsman with a boxing blue, a combination that made him an excellent soldier. After Sandhurst, he had served a number of tours in Northern Ireland until in 1977, when he was based at Portadown, spending several nights each week at the SAS base in Bessbrook Mill in County Armagh. It was widely rumored at the time that he was a member of the SAS. In fact he was never a member of the SAS; he was seconded to the army's undercover intelligence gathering unit, known as 14 Intelligence and Security. However, he did work closely with the SAS, gathering information for men from the Regiment to act upon.

Nairac was convinced that he could pass himself off as an Irishman, speaking with an Irish accent that was good, but not perfect. He had developed a Belfast accent which he could speak fluently and enjoyed going to local pubs in his brown Triumph Dolomite to chat to the locals and join in the evening singsongs. He went by the name of Danny Boy, a name by which many SAS soldier called him. However it is difficult to pass yourself off as an Irishman and thus get accepted by the local community. Although in his favor Nairac did have bravado, which to some degree helped him survive for a time, anyone who is willing to get up and sing songs in the local bars of Bandit country comes under scrutiny, and it would be an insane gesture by any member of the security forces to act so, especially alone. Many of the locals liked Nairac and those that had anything to do with the terrorist organizations believed him to be some form of plant working for the Official IRA; these plants were known as a "Sticky."

In fairness he did gather much basic information but his method was crude and dangerous. Moreover many think he enjoyed the idea of being a lone undercover agent. This may seem an unfair comment but he did forgo standard operating procedure

Rare image of Captain Robert (Bob) Nairac dressed in undercover clothing. A very brave man.

for any covert operation—often working without adequate backup. On the night of May 14, Nairac went to the Three Steps Inn at Drumintee in County Armagh. The pub is isolated, on a lonely hill out of range of any immediate help, and three miles from the border of the Republic. A couple of days prior he purchased the song sheets to two well-known Republican songs and practiced them in Bessbrook mill until he knew them by heart. Feeling confident, he left the barracks at around 7:30 p.m., making his way to the Three Steps Inn—he told no one of his plans or where he was going, only arranging to telephone the SAS desk operator at around 11:30 p.m. The pub was packed that night with around two hundred people, and after an evening of steady drinking the songs began. Nairac had been on stage and finished his version of two favorite IRA tunes, "The Broad Black Brimmer" and "The Boys of the Old Brigade."

Eventually he decided he had had enough and made his way out to the parking lot, where he was followed and questioned as to his identity. A fight started; Nairac should have been able to handle it as he was a keen boxer and could easily take care of himself. However, while the fists were flying, his Browning 9mm pistol fell onto the ground. This was grabbed by one of his attackers and he was quickly overpowered and knocked unconscious. His captors bundled him into a car and immediately took him south over the nearby border. For a short time he was left unconscious in a house with only a single guard. Nairac recovered and attacked his guard, knocking him down. He grabbed the guard's revolver just as another IRA man rushed into the room, pointed the pistol at him, and pulled the trigger. The weapon misfired. Nairac pulled the trigger again and it misfired once more. He pulled the trigger a third time and once again it misfired. By then the original guard he had attacked moved in from behind and knocked him unconscious. His internment at the house had given time for an IRA unit to be called in after which his original captors were dismissed. Nairac was swiftly moved from the house and taken into Ravensdale Forest near the border. He was bundled out of the car and carried into a small field beside a bridge. For several hours he was brutally tortured by his captors who wanted to find out details of current SAS operations. The interrogation took the form of a severe battering about the head and body with a fence post. Despite the torture, Nairac refused to speak. Knowing he was about to die and being a staunch Catholic, he asked his captors for a priest. In a final and macabre humiliation, one of the terrorists played the part of the priest to hear the officer's confession. Thereafter, realizing they would not get anything out of him, his IRA captors shot him with his own pistol. Because of his capture, Nairac failed to make his 11:30 p.m. call back to base at Bessbrook, and due to his maverick, loner attitude this was not thought to be abnormal and thus the alarm was not raised until 6 a.m., by which time he was already dead. SAS units were deployed to search the area but without any success. Two days later, the IRA issued a statement saying that he had been killed. His body was never recovered.

Author's Note: Some years later I was privy to information as to the disposal of his body when in 1977, a twenty-four-year-old joiner from Meigh in County Armagh, Liam Patrick Townson, was arrested by the Irish police on suspicion of being involved in Nairac's murder. Under close questioning, Townson gave an accurate account of Nairac's interrogation; the methods used were so horrific that it has never been made public, but suffice to say that it involved a strong rumor about a certain animal-feed processing plant. Townson also drew a detailed sketch of the scene of the killing, pointing out where the terrorists had hidden Nairac's Browning and clothing in the nearby forest. Strangely enough, as Townson related the story, he had nothing but the greatest respect for the way in which Bob Nairac suffered in silence. Townson and five other men from the north were jailed for life for their part in Nairac's killing. Two months later, in February 1978, Bob Nairac was awarded the George Cross, the highest peacetime honor a serviceman can receive. As a result of his death, new procedures were put in force by the SAS with auto alert transponders being issued in order to locate any SAS member who may find themselves a captive of the IRA.

Nairac's name has surfaced several times since his death. In 1984, seven years after his death, allegations arose that linked him to the 1975 murder of John Francis Green, an IRA commander in North Armagh, who was shot dead at a remote mountainside farmhouse in County Monaghan in the Republic. Other allegations led some to believe that Nairac was actively assassinating terrorist suspects and therefore supporting the idea that the British had a shoot to kill policy. This is disputed by those who worked with him and backed-up by the fact that although armed on the night of his capture, and recognizing that his original assailants were not IRA, he did not use his pistol.

TCG (TACTICAL CONTROL GROUP)

TCG units were set up in Northern Ireland to coordinate covert and overt police and military operations. The main aim of TCG was to best unitize the various sources of information available and task the appropriate agency to carry out any resulting operation. The province had three TCG units positioned in Belfast, Londonderry, and Armagh, each one controlled by the RUC Special Branch. Each unit also had liaison officers from CID, SAS, 14 Int together with Brigade intelligence officers, who would answer to the senior Special Branch officer in charge of TCG. In practice the units worked extremely well, controlling both the flow of high-grade intelligence and the movement of troops on the ground, which avoided friendly forces contact. Many of the successes against the IRA in Northern Ireland can be contributed to the controls provided by TCG.

E4A

A unit known as E4A was an intelligence gathering unit set up in 1978 using personnel from E Department of the RUC. It was primarily made up of male and female officers from the Special Branch. Initial training was provided by the SAS in camp 1 in Herefordshire. The unit operated closely with both the SAS and 14 Int & Security all under the control of the RUC Special Branch TCG Headquarters.

IRA bombing of Loughgall Police Station, the van can be seen in the bottom left hand corner.

TYPICAL SAS OPERATION IN NORTHERN IRELAND

The following is a classic SAS operation, which took place in Loughgall, Northern Ireland, on Friday, May 8, 1987. Intelligence had been received to indicate that the RUC police station at Loughgall was to be attacked in the method used the year before in County Armagh. That incident had taken place in April 1986, when a mechanical digger had been packed with explosives and driven into the RUC station at the Birches. It had caused widespread damage. A report that another JCB had been stolen in East Tyrone gave rise to the suspicion that an identical IRA operation was being planned. All efforts were made to locate the digger and identify the target. After intensive covert searching by both SAS, E4A, and 14 Int & Security units, the weapons and explosives were located. Subsequently, the digger was also located in a derelict building on a farm some 15km (about 9.3 miles) away. Surveillance by E4A provided more information and eventually the target was identified as the RUC station at Loughgall. This station was only manned part-time and consisted of one principle building running parallel to the main road, surrounded by a high wire fence. The time and date of the attack was eventually confirmed through a Special Branch telephone tap.

Two of the IRA activists were named as Patrick Kelly and Jim Lynagh, who commanded the East Tyrone active service unit. When masked men stole a Toyota van from Dungannon, Jim Lynagh was spotted in the town, suggesting that the van was to be used in the Loughgall attack. Not long after, the OP reported that the JCB was being moved from the derelict farm. At this stage the SAS, who had been reinforced from Hereford, took up their ambush positions. It was reported that some were in the police station itself, but this was not true—instead most of the main ambush party was hiding in a row of small fir trees which lined the fence on the opposite side to the station. Several heavily armed stops were also in position, covering all avenues of escape.

At a little past 7 p.m., the blue Toyota van drove down the road in front of the police station. Several people were seen to be inside. A short time later it returned from the direction of Portadown—this time followed by the JCB carrying three hooded IRA terrorists in the cab. Declan Arthurs was driving, with Michael Gormley and Gerald O'Callaghan riding shotgun. The bucket was filled with explosives contained in an oil drum, which had been partly concealed with rubble. As the blue van charged past the station, the JCB slammed through the gate. One of the two terrorists riding shotgun (although it was not clear which one at the time) ignited the bomb and all three made a run for it. Back at the van, several hooded men jumped clear and started to open fire in the direction of the RUC station. At this stage the SAS ambush was activated.

The sudden hail of SAS fire was terrifying. All eight members of the IRA fell under the hail of bullets. At the height of the firefight, the bomb exploded, taking with it half the RUC station and scattering debris over all concerned. As the dust settled, the SAS closed in on the bodies. At that moment, a white car entered the ambush area with its two occupants dressed in blue boiler suits similar to those worn by the IRA. They were unfortunately mistaken for terrorists—especially when, on seeing the ambush in progress, they stopped and started to reverse. One of the SAS stops opened fire, killing one of the occupants and wounding the other. It later transpired that the dead motorist, Antony Hughes, had nothing to do with the IRA. Several other vehicles and pedestrians soon appeared on the scene, but by this time the situation was stabilized.

The van used by the IRA hit team at Loughgall, the SAS were hidden behind the fence no more than two meters away, hence the devastating carnage.

Loughgall was one of the most successful operations ever mounted against the IRA, who were totally stunned by the loss of two complete active service cells. The Hughes family was compensated for their loss, and with no public inquest the matter was closed. The IRA, believing that there was a mole in their organization, went into a period of self-assessment, but they did not lick their wounds for long. Shortly after, on Remembrance Day, at a ceremony in Enniskillen, the IRA detonated a massive bomb. Eleven were killed and more than sixty were injured.

The 1980s did not start well for the SAS, especially for those working in Northern Ireland. The SAS continued to operate in conjunction with the regional Tactical Control Group (TCG). As always, the Special Branch of the Royal Ulster Constabulary (RUC) would get their hands on certain information which required the hard-hitting skills of the SAS. Unfortunately, Special Branch was always very tight-lipped on where or how the initial information materialized. While this was good security, it did not help those troops who carried out the operation.

In any event, Special Branch (TCG) had requested that the SAS Armagh unit take on three separate tasks. Unbeknownst to the SAS, the information that triggered these operations came from a reliable telephone tap. The first task was to watch two part-time policemen whom the IRA planned to assassinate. The information from the TCG clearly stated that they would be hit somewhere close to the home they both shared. This was straightforward work for the SAS, who inserted an observation post (OP) close to the home of the two policemen. All they had to do was sit and wait for the IRA to make their move.

Bombed public house similar to the one the SAS left temporarily unprotected.

The second task was to protect a public house which the IRA had threatened to burn to the ground. Again this was a simple operation which once again involved inserting an OP and waiting for the IRA to turn up. Logistically, servicing two OPs, with resupply and the changing over of men was not too much of a problem until the weekend came around, at which time TCG had a priority operation which required every man in the field.

It would seem that one of the local underlings in the IRA had seriously upset the organization. The sentence for his crime was to be kneecapped, with the punishment being carried out by a leading IRA member. The TCG were so desperate to catch such a leading IRA figure in the act and employed a massive surveillance operation each weekend. This involved the SAS, E4A, and 14 Intelligence Unit in a huge stakeout of a particular Republican haunt which ran a disco every Saturday night. The problem for the SAS was one of manpower, and in order to commit themselves to the weekend operation, they were forced to lift the men working the OP protecting the public house every Saturday night.

On week two the guys in the OP watching the home of the two policemen were spotted by a local farmer and asked to be withdrawn. TCG were informed of the predicament and begged the SAS unit to stay in position as a hit on the two policemen was imminent. However, the final decision was down to the men on the ground, who, for safety, opted to withdraw. Twelve hours later, when the two police officers were returning from duty, their car was badly shot up. The IRA had positioned themselves between the house and where the SAS OP had been; had the SAS remained in place they would have had the IRA gunmen. As luck would have it, only one of the two policemen was injured in the attack. TCG were not pleased and while the senior SAS man had made the correct decision he was severely chastised.

Author's Note: Had the SAS OP known the intelligence of a wiretap was coming, they would have risked staying in position. It was this lack of cooperation and intelligence-sharing which remained the source of all the problems.

The following weekend every available man was committed to the operation around the disco, and as usual the OP on the pub was pulled off for the night. The surveillance operation went smoothly, but as with the previous two weekends, nothing happened. At around two in the morning, the operation was called off. One SAS car was to drop off the crew for the Pub OP on its way back to base. They approached the public house only to find it ablaze: the IRA had done their job during the brief moment the OP had been lifted. The SAS refer to this as the "Paddy Factor." Needless to say by now TCG were extremely upset.

It was not just the SAS unit in South Armagh that was having bad luck; the Belfast unit was also getting its share. In early April a number of IRA weapons were discovered. Although these had originally been under surveillance, the IRA had managed to move them to an unidentified hiding place. Among these weapons was a 7.62mm M60 machine gun, which had originally been stolen in America. It was vital that the weapons be relocated. Luckily some of the weapons had been fitted with sophisticated location devices. While this beacon only had a short range and limited battery power, after an intense search a weak signal was discovered, locating the weapons in a terraced house in Antrim

Road. However the signal was so weak the exact location could not be identified and it could have been in any one of three houses. After an intelligence check by the RUC Special Branch, it became known that the IRA had previously used one of the three houses. Working on the best bet principle, the SAS mounted an operation to recover the weapons.

Angelo Fusco, one of the IRA "M60 Gang."

On the afternoon of May 2, 1980, two cars containing SAS soldiers headed down the Antrim road, screeching to a stop outside house number 367. Another vehicle containing three more SAS soldiers secured the rear. For security reasons, there had been no cordon or military activity prior to the raid, and the SAS team stormed straight at the house. Unknown to them, the IRA had mounted the missing American M60 in the upstairs window of the adjoining house, number 369. Captain Westmacott, who had been sitting in the middle of the rear seat and was the last to move, caught the full blast as the M60 opened fire. He was killed instantly. Realizing what had just happened, the whole assault was quickly switched, but by this time the IRA man had surrendered. The sound of gunfire brought the Army and RUC to the scene; additionally a Catholic priest suddenly materialized to see that the IRA man was allowed to surrender. One of the men in the house was Angelo Fusco—although Italian, Angelo joined the west Belfast brigade of the IRA and was a member of the notorious M60 gang. On June 10, 1981, Fusco and seven other prisoners took a prison officer hostage at gunpoint in Crumlin Road jail. They overpowered other officers and solicitors visiting the jail and made good on their escape to waiting cars. Fusco was convicted in his absence to life imprisonment but was granted a Royal pardon in 2014 and allowed to return to Northern Ireland.

Captain Richard Westmacott.

The tragic events of May 2 caused the SAS to drastically rethink their tactics, and today, in such a situation, the house would be put under surveillance before they would allow any direct intervention. Captain Richard Westmacott was the first SAS soldier to die in Northern Ireland, having joined the Regiment from the Grenadier Guards and serving as a Captain in G Squadron. With curly fair hair, schoolboy looks, and a love of poetry, his appearance belied his performance. He was a remarkably tough member of the SAS and in recognition of his achievements, was awarded the posthumous Military Cross.

Author's Note: Captain Westmacott's cousin is Sir Peter Westmacott, a British Ambassador who facilitated the first meeting between Gerry Adams and Sir Patrick Mayhew which helped bring the troubles to an end. I had the pleasure of meeting Sir Peter at the embassy in Washington last year.

Nevertheless, Westmacott's death seemed to end a run of bad luck for the SAS. A few days later, the anti-terrorist team successfully stormed the Iranian Embassy in London. The anti-terrorist unit had been in operation for over five years and at last had the opportunity to show what it could do. Some team members had been present when the Dutch Marines had stormed into the train that had been taken over by Molucan terrorists, while others had assisted the GSG9 throughout the Mogadishu hijacking. But now the SAS were to get their own venue and they were to make it a real classic; moreover, it was filmed on national television and beamed around the world.

DRUMNAKILLY

Another operation that shows insight into SAS Operations was the killing of three IRA members near Drumnakilly. The operation took place on August 30, 1988, which resulted in the death of three members of the Mid-Tyrone Brigade. Martin and Gerard Harte and Brian Cullen were suspected of being responsible for over thirty-two murders, but none had ever been proved and the three remained at large.

OPERATIONAL PLAN

Two items came to light which helped formulate the plan; first intelligence showed that the Mid-Tyrone Brigade planned to shoot a retired Ulster Defense Regiment (UDR) officer, living and working in the Bandit Country of Omagh. Around the same time, the SAS had also uncovered a weapons dump nearby—which was believed to be a cache of personal weapons belonging to Cullen and the Hartes.

It was decided to set a trap to capture both Cullen and Hartes, given that they knew the target and the location of the weapons; it was a simple matter of enticing the IRA members into a trap.

An OP was set up to watch the weapons cache for any activity, and the UDR officer was given a Leyland truck with an armored cab and advised to establish a predictable daily routine on his way to and from work along a quiet country road. The plan was to spur the terrorists into recovering their weapons, when the SAS would be alerted that the operation was on. The UDR man was to be substituted by an SAS man and the truck designed to break down in a designated place, thus setting the ambush.

However, before any of this could take place, the Mid-Tyrone Brigade attacked another target. On the afternoon of August 20, 1988, a bomb exploded in a small trailer parked by the side of the road just as a bus carrying soldiers of the 1st Battalion Light Infantry was returning to the Omagh Barracks. Twenty-two pounds of Semtex threw the bus into the air, killing eight soldiers, injuring twenty-seven others, and creating a six-foot-deep crater in the road. This victory for the IRA, however, was to be short-lived.

Nine days later, the OP at the weapons cache saw the terrorists approach to collect their weapons. This message was relayed immediately, assuming that the Mid-Tyrone Brigade was planning the murder of the Ulster Defense Regiment man the next morning. Shortly before dawn, three SAS soldiers set out on foot to the location of the planned ambush, a deserted farm just outside the village of Drumnakilly, where they hid and waited. Another SAS soldier, dressed in the UDR man's overalls, began the drive along the routine route until the truck "broke down" near to the farmhouse, when the soldier got out to try and mend it. Unbelievably, it took six hours for the news to reach the terrorists, who eventually turned up in a stolen car but not the one that had been expected. The SAS driver, masquerading as the UDR man, was caught unprotected as the terrorists, dressed in the standard uniform of blue boiler suits and black balaclavas, began to fire at him. He managed to use a brick gatepost as cover, which allowed the other SAS soldiers a clear view of the Mid-Tyrone Brigade. Seeing their target hide behind the gatepost, the terrorists got out of the vehicle to complete their task but were killed outright by unrelenting firepower from the SAS soldiers in their hideouts. The successful operation was completed when a lynx helicopter arrived to return the soldiers to their base within minutes of the shooting—even before the press arrived. The whole exercise was deemed a great success for both the SAS and the Security Services in Northern Ireland.

COALISLAND SHOOTING

On Sunday, December 4, 1983, an SAS OP, which had been watching an IRA arms cache, opened fire on two men who arrived to collect the weapons with the intention of committing a murder. The weapons were housed in a hide protected by a thick hedgerow that separated a small field from a quiet country lane—a gateway into the field providing easy access. Once the hide was discovered, a full-scale operation was mounted; this involved 14 Intelligence and Surveillance group who were responsible for watching the IRA suspect players, the SAS would act as cover for the weapons, and E4A who would form a Quick Reaction Force (QRF). The SAS inserted several two-man OPs around the field, with the main OP in a ditch opposite the weapons hide, some twenty meters away. E4A were positioned at Lurgan while 14 Int carried out tight surveillance.

The SAS men in the OPs remained in position despite the December weather, which was wet and cold. At around 3:15 p.m. on Sunday afternoon, a car was heard slowing down and eventually stopping opposite the gate. Two men got out and climbed the gate, leaving the driver in the car. They walked directly to the hide, knelt down, and retrieved the weapons. The first terrorist, Colm McGirr, pulled out a weapon and passed it to Brian Campbell. As Campbell stood, turning back in the direction of the car, the two SAS soldiers challenged from the OP. McGirr, who was still kneeling by the bush, turned with a gun in his hand and was shot dead. Campbell ran for the gate still holding the weapon. Two more shots rang out and he fell, mortally wounded. The driver, on hearing the gunfire and realizing that it was a trap, drove off. An SAS soldier jumped out to stop him, firing four high velocity rounds directly into the car but failed to stop it.

Two of the SAS operators in Northern Ireland OP—the one on the right was the first soldier on the scene after the Coalisland shooting.

A quick check indicated that McGir was dead, but Campbell was still breathing and an SAS medic inserted an airway and administered first aid; Campbell died before the ambulance crew arrived some twenty minutes later. The area was sealed off by E4A, who were activated via radio instructions to search for the missing getaway car. It was quickly located

near some houses about a mile away, but despite a great deal of blood at the scene, the injured driver was not found. It was later learned that he had been whisked away by local Republican sympathizers who managed to get him across the border into the Irish Republic before major stops came into force. In the eyes of the SAS it was a clean, neat job, apart from the vehicle stop, for which the SAS soldier concerned was severely criticized by the regimental head-shed for allowing the car to pass.

"FLAVIUS" OPERATION

Late in 1987, Sean Savage, an infamous IRA bomb maker, was discovered living in Spain with another IRA suspect, Daniel McCann. For six months, MI5 kept the two under surveillance, convinced that they were planning further bombings.

Their suspicions were further increased when Maraid Farrell arrived at Malaga airport on March 4, 1988, where she was met by Savage and McCann. MI5 continued to keep a close watch on the three, later acknowledged by the IRA to be an active cell, recording several of their conversations. Before long they discovered that the trio was planning to attack the British Garrison in Gibraltar, with a car bomb at a military ceremony, where several regimental bands would be parading.

The Gibraltar police were informed and the SAS requested to send in an anti-terrorist team, under the operational name of "Flavius," with instructions that the IRA active service unit was to be seized. Although contact with the IRA cell was lost for a short while, the SAS believed they understood the IRA's plan. They suspected that a car, designed as a decoy, would be delivered to Gibraltar and parked in a prominent position along the planned route for the military parade, thus guaranteeing a parking space for the actual car bomb. Most of the troops and public would assemble in the plaza, which the SAS correctly assumed to be the IRA's target area since it was obvious that a bomb here would cause the greatest destruction. At 2 p.m. on March 5, Savage was seen parking a white Renault 5 and was thought to be setting up the bomb triggering device. At about the same time, Farrell and McCann were observed crossing the Spanish border, making their way into town.

The bodies of McCann and Farrell shot dead by the SAS in Gibraltar.

As soon as Savage left the scene, the SAS men were called into action, using an explosives expert to undertake a walk-past of the Renault. Although there were no obvious signs of a bomb, such as the rear suspension being depressed, it was agreed that the car probably did contain an explosive device—possibly Semtex—which is relatively easy to conceal. Joseph Canepa, the local Police Chief, immediately passed control to the SAS, whose orders were to capture the three IRA members alive, if at all possible. However, as in all such situations where lives may be at risk, the SAS have the right to shoot to kill.

Each SAS soldier was dressed in casual clothes and was armed with 9mm Browning Hi-Power, able to maintain contact via small radios hidden on their person. Meeting up with McCann and Farrell, all three IRA suspects made their way back towards the Spanish border, closely followed by four of the SAS team. The team was forced to split into two, however, when Savage, for no apparent reason, turned round and began to make his way back into the town.

A few moments later a local policeman was recalled to his police station, just as he happened to drive past McCann and Farrell, causing them to turn nervously. McCann instantly made eye contact with one of the SAS soldiers, who started to issue a challenge, being no more than ten meters away from the suspects. McCann immediately moved his arm across his body; fearing he was about to remotely detonate the bomb, the soldier shot him in the back. Farrell, it is said, distinctly went for her bag and was shot with a single round. By this time the second soldier had also opened fire, hitting both terrorists. Savage heard the shots and turned back, only to be confronted by the remaining two SAS soldiers, who shouted a warning to him. Savage ignored the warning and tried to reach what the soldiers took to be a weapon. Both SAS men fired and Savage was killed outright.

At first the mission was hailed as an outstanding success, but the euphoria was short-lived. All three terrorists were found to have been unarmed, and while a bomb was later discovered in Malaga, one was never found in the suspects' car. Allegations were made and accusations flew from "witnesses" who claimed to have seen the whole episode. They claimed that the three had surrendered with their arms in the air but had been shot at point-blank range, while they lay on the ground, by the "unscrupulous" SAS, living up to their shoot to kill policy.

A two-week inquest followed in September 1988, at the end of which the jury returned a majority verdict of lawful killing. Outraged by the outcome, relatives of the three IRA members took the SAS soldiers responsible for the killing to court—although the European Commission of Human Rights in Strasbourg decided, 11 to 6, that the SAS did not use unnecessary force. It was decreed that the soldiers were justified in opening fire given the circumstances. However, they did refer the case to the European Court of Human Rights. As a result, the British government was forced to pay compensation.

SUMMARY

When the SAS were first deployed to Northern Ireland, it was to fulfill a specialist one-off mission, but it was not long before a whole Squadron was deployed into the province. In the early days it was a simple matter of getting to know the country and the problem. More importantly, these were our own people and while the Protestants and the Catholics may have been at each other's throats, as soldiers from mainland Britain we did not truly understand this grievance and division. So it was that the tactics and skills required took some years to fall into place.

DEPLOYMENT

In the very early days the Squadron would live and operate from one location, acting more or less as infantry soldiers. However, as time passed and the problems of Northern Ireland were better understood, the Squadron split into three groups with a separate HQ which allowed for better deployment and coverage of the province.

Operations for all three sub-units were more or less the same, although those stationed to cover Belfast rapidly converted from military clothing and infantry tactics to a more civilian and surveillance role. Those troops operating in the North and South remained operating as infantry for a longer period.

DRESS

Dress was very much dependant on the particular operation; if you were in an urban OP then you wore civilian clothing; if in a rural OP you wore military camouflage dress. There were the odd exceptions but only when operational conditions demanded.

Hair was grown longer, and you were not required to shave when operating in civilian clothing. Most of the latter would come from the local secondhand shops or purchased from the markets—the idea being that the SAS operators looked like every other civilian walking around.

Author's Note: Within the SAS Squadrons were several men from ethnic backgrounds whose skin was either brown or black; it was a simple matter of recognition that these men could not be used on covert operations during daylight hours in civilian clothing as there were very few people of color in Northern Ireland.

TACTICS & SKILLS

The tactics in Northern Ireland varied greatly from straightforward infantry patrols to highly complicated surveillance roles. Over the years and as the SAS evolved, there became a set pattern to local operations. The establishment of TCG proved a great asset to the SAS as tasking became more organized. Special Branch would identify a particular requirement based on either physical information such as the location of a covert bomb or intelligence from a telephone tap. Either way, the local SAS liaison officer would be given the task to study and then make recommendations in coordination with the other agencies. If the information and intelligence looked sound, then the liaison officer would task the SAS troops operation in that area.

Based on the type of intelligence, the SAS would allocate a number of men had to first confirm the intelligence or information, and then act accordingly. For example, if a culvert bomb was found under a road waiting to blow up a passing police or military vehicle, the first thing would be to locate the trigger end; secondly to deactivate the bomb. This latter task was always done by the Ammunitions Technical Officer (ATO) escorted into the site by the SAS team—most of which stood well back.

Much of the work involved in setting up an OP which could, at best, take pictures of the suspects and observe their movements. In a rural environment this was fairly simple and access to extremely long-range cameras allowed the OP position to be some distance from the target. Being too close in the countryside would inevitably cause the dogs to bark nonstop and expose your presence.

Urban OPs were much more difficult, as setting up in an area where people live requires a lot of skill. First the best location is found by means of a CTR. Next is the actual building penetration; if this was in a public building it was fairly easy, but getting into the attic of an occupied house is not so stress-free.

Author's Note: I once managed to climb into an attic via a skylight in a row of terraced houses. With care, a fellow soldier and I made our way from attic to attic until we reached the desired house. At this stage I removed a trap door in the ceiling and lowered myself into the bedroom, only to find an old lady asleep. I managed to extract myself silently and look for another position for the OP, which was in the roof of the local bank.

From time to time, surveillance operations and hard assaults on buildings require the use of covert cars. The cars were also used for nightly drop-off and pickup of the various men out on operations. In general, each of the three detachments in the province had around twenty cars and a couple of vans for deployment.

Skills-wise, there was the need to learn the various codes; these included spot codes for the cars traveling around the province. The driver would simply say into his hidden microphone "towards RED 3," and the desk operator would confirm and make note of the vehicle's position on a large map in the operations room. This way the location of all vehicles and personnel out on operations could be monitored.

Author's Note: There was no GPS at this time and it was not until 1989 that the first Block II satellites were launched. This program finished mid-summer 1993 and completed the modern GPS constellation of satellites that we know today.

Lock picking was another skill required; in this field the SAS excelled, producing some of the best locksmiths in the world. When a lock or door could not be opened, then such items as hydraulic rams were used to spread the doorposts and literally pull the bolt out of the lock. Releasing the pressure would allow the door to close again.

WEAPONS

Initially, it was thought that normal infantry weapons such as the SLR would be best suited when the SAS were patrolling in combat clothing, as carrying an AR15 would only pick them out as Special Forces. However, as the roles changed AR15 /16 became ever more popular especially when patrolling n at night.

Covert weapons included the browning 9mm pistol and the Walther PPK, as this small pistol was easy to hide. When working in confined areas it was normal to use the Irgam MAC 10—a one-handed chin pistol with a lot of firepower.

Chapter 8

SURVEILLANCE & INTELLIGENCE GATHERING

While the SAS were originally conceived as a raiding force operating deep behind the German lines of World War II, they soon learned that gaining good intelligence would result in decisive victories during operations. The early cross-border "Claret" raids into Indonesia saw its birth in the basic form, but since that time the SAS has become legendary when it comes to surveillance and intelligence gathering. However, their acceptability into the murky world of espionage was won the hard way.

Although Britain's intelligence and security community is politically complex, in real terms the politicians have little real authority over the activities and operations of a secret nature. The UK Intelligence Community consists of the Secret Intelligence Service (SIS) or MI6 (overseas), MI5 (domestic), and Government Communication Headquarters (GCHQ). There are many other units such as SIGINT and COMSEC agency and the Defense Intelligence Staff (DIS), plus Special Forces. Then there are the oddballs such as Diplomatic Telecommunications Maintenance Service (DTMS), which provide bugging and debugging services with experts to sweep sensitive government facilities.

For many years the SAS kept their own council when it came to surveillance and intelligence gathering. Their techniques were honed during their many different types of operations around the world. Over the years the SAS has grown stronger mainly due to its confidence and ability to get the job done with as little fuss as possible. These skills and abilities did not go unnoticed, but it was the decline of the Special Political Action Section and the birth of the SAS Counter Revolutionary Warfare Wing, that started the first engagements with SIS.

SPECIAL POLITICAL ACTION SECTION (SPA)

This unit was to become heavily involved in deception, political influence operations, and engineering changes in the leadership of foreign countries through rebellions such as the failed Indonesian Permesta rising of the late 1950s, insurgency, coups such as Iran in 1953 and the Congo in 1961, or perhaps by extreme executive action. Within SIS the SPA was known as the "jolly fun tricks department," and was directly controlled by the Head of R1 (Political Foreign Office). The section used many assets ranging from foreign mercenaries, former intelligence or the odd Special Forces personnel, in what was known as deniable operations. The SPA was quietly abolished due to the Labor Government's aversion to covert action in the mid 1970s. It was at this stage that the relationship with the SAS, seconded and retired, as well as a number of private specialist companies, became ever more important; and by 1987, a Special Forces Directorate was formed to coordinate the activities of the SAS and SBS and to ensure closer collaboration with the SIS. By 2003, such activities were the responsibility of the General Support Branch.

General Support Branch (Known as the Increment)

The General Support Branch is a small number of Special Forces soldiers attached to SIS who are known as the increment and who carry out dirty, deniable operations. Personnel for these covert actions are provided by the 22 SAS CRW—Counter-Revolutionary Wing and M (CT) Troop of the SBS and are supported by the RAF S & D Flights. The SAS CRW Increment would normally have around ten to fifteen highly trained specialists available for SIS requirements. All have a minimum of five years service in the SAS, are of the rank of sergeant or above, and have been heavily vetted by SIS. They will have gone through an induction course on surveillance and intelligence as well as three weeks at Fort Monckton, the Army's secretive FRU (agent running unit). The 14th Intelligence & Security Company, now renamed SRR, provide surveillance for the SAS and SIS.

Today the SIS also directly employs a considerable number of former members of the SAS, SBS, and SRR. These retired SAS Special Forces carry out covert tasks around the world. Their previous operational experience, skills, and intelligence makes them ideal candidates for espionage.

UKN

Little is known about UKN other than that SIS employ up to fifty or so United Kingdom Nationals, either retired Special Forces who are doing business on a worldwide scale or who work for a specific company selling arms or related services overseas. UKN often work alongside the increment.

CRW provided a large number of courses including bodyguard training and VIP protection programs.

COUNTER REVOLUTIONARY WING

The Counter Revolutionary Warfare Unit (CRW) within the Regiment can trace its origins back to the Keeni Meeni operations in Aden. In early 1970, the regiment formed its own CRW unit, with the special purpose of developing techniques to counter terrorism. From its inception, CRW was a vision of how the modern day SAS soldier was to develop. No longer would they always fight in military clothing; CRW tasking called for civilian clothing and a variety of new covert weaponry. Their brief was to study the growing terrorist threat and come up with ways of combating the problem. CRW continued to grow and soon became the hub of all anti-terrorist training, techniques, and methods. All types of surveillance were practiced: vehicle, foot, and observation methods. Methods of entry and the skillful use of explosives were honed. Good camera skills were learned by the SAS, as were the means to develop and print the film; even infrared photography was taught for night work. Close quarter battle (CQB) skills including advanced shooting techniques and self-defense combinations with bodyguard courses (BG) were organized. VIP protection drills were learned, and SAS personnel were sent on police high-speed driving courses. In Hereford itself, a purpose-made building was constructed which was to become known as the killing house. This training facility became famous around the world for its unique features: SAS members could practice shooting inside an aircraft, snatching a hostage from captives, or simply taking out a terrorist target. It was a time of change, and the number of SAS men turning up for daily work in civilian clothes increased dramatically.

Today the duties of the SAS CRW unit span the world. They infiltrate into enemy territory, gather intelligence, support our allies, undertake demolition work, provide bodyguards for VIPs, and rescue hostages—plus numerous tasks we never read about.

SPECIAL RECONNAISSANCE REGIMENT (SRR)

The Special Reconnaissance Regiment (SRR) was established on April 6, 2005, and is part of the United Kingdom's Special Forces, which is under command of Director Special Forces. The SRR can trace its history back to the early 1970s with the formation of the Mobile Reconnaissance Force (MRF), which after being compromised by the Irish Republican Army (IRA), subsequently became 14 Intelligence Company (14 INT), or "The Det." The skills learned fighting the IRA were deemed too useful to simply abandon and thus the SRR was conceived. The SAS and SBS rely heavily on reconnaissance and intelligence provided by SRR.

Stationed in Credenhill, Hereford, with the SAS, the new unit is small with around ninety operators, but unlike the other Special Forces units, SRR trains both men and women. The role of the Special Reconnaissance Regiment, as its name implies, is to conduct reconnaissance operations where and when required, thus relieving other UKSF units of surveillance duties and allowing them to concentrate on offensive operations.

All members of the SRR are drawn from the British Armed Forces with selection taking place before applicants attend a course at Hereford. Surveillance requires many skills such as advanced driving, CQB, and the art of remaining invisible. Most of these skills are developed in the Pontrilas area a few miles south of Hereford. Resistance to interrogation is also a must; training for this takes place with the Joint Services Interrogation Unit in Ashford, Kent.

The role of the SRR is not limited to the Special Forces; when required the SRR will also carry out work for MI5 and MI6. In March 2009, they were known to have ventured back into Northern Ireland to gather intelligence on dissident Republicans. The SRR were also implicated in the surveillance of Jean Charles de Menezes before he was shot by police marksmen.

Author's Note: Lance-Corporal David Jones was a member of 14 Intelligence & Security Unit enlisted from the Parachute Regiment, who was killed in Northern Ireland on March 17, 1978. He and another member of the unit were mounting a covert surveillance operation near the village of Maghera, County Londonderry. Three men approached them in combat dress, and Jones, thinking they were members of the Ulster Defence Regiment, stood up and issued a warning. The men, who were terrorists, acted fast and Jones was unable to protect himself as they fired, hitting him in the stomach and also wounding his companion. However, Jones did manage to fire his weapon, wounding one of the terrorists before he fell. The wounded terrorist was eventually captured after a long search and was identified as Francis Hughes, a member of the Irish Republican Army. Jones died from his wounds the day after the attack.

The whole unfortunate incident meant that lessons had to be learned, not only in communications between the army, police, and SAS, but also that operatives needed to be more cautious and react more quickly in such situations. All of these problems were soon addressed and rectified. David Jones was a good friend of mine.

18 SIGNALS REGIMENT

The 18 Signals Regiment (UKSF) was formed in 2005 to encompass all existing signals units, such as 264 Signals Squadron (SAS). The idea was to form all the smaller sub-units of the SAS, SBS, SRR, and SFSG into one Signals' regiment which would come under control of Director Special Forces (DSF).

In addition to the amalgamation, there was also a need to improve communications and IT capabilities. These included new satellite equipment for communications and tracking of individuals, both via GPS location and direct satellite imagery, the latter being focused initially on Iran and from 2007 directed towards Afghanistan.

The 18 Signals Regiment also has a wide range of monitoring equipment including the ability to tap into most mobile phones and thus locate the nearest network cell tower. This is particularly handy for locating known terrorists and for guiding SAS teams onto target.

Anyone volunteering for the 18 Signals Regiment must undergo a form of selection which, while not as difficult as the SAS selection, is still extremely hard. Those who pass the selection are expected to accompany regular SAS Squadrons into theatre, and therefore skills such as conduct after capture must be learned. The course starts with a technical trade assessment which lasts about one week, followed by a six-week course of SAS support communications, physical training, weapons, etc. Finally all members of 18 Signals' Regiment go through the All Arms Basic Parachute Course. The 18 (UKSF) Signals Regiment consists of the following units:

- 264 Signal Squadron is dedicated to 22 Special Air Service Regiment
- 267 (SRR) Signal Squadron is dedicated to The Special Reconnaissance Regiment
- 268 (SFSG) Signal Squadron is dedicated to The Special Forces Support Group
- SBS Signal Squadron (Royal Marines) and is dedicated to The Special Boat Service
- 63 (SAS) Signal Squadron (V) are a Royal Signals Territorial Army unit dedicated to both SAS Reserve Regiments, 21 and 23 Special Air Service Regiments.

SPECIAL FORCES SUPPORT GROUP (SFSG)

On December 16, 2004, the Secretary for Defense announced to Parliament that the government intended to set up a new unit in support of the SAS in the global fight against terrorism. The unit was to be based at the RAF base in St. Athens, near Cardiff South Wales. Personnel for the new unit were drawn from the Parachute Regiment, the Royal Marines, and the Royal Air Force Regiment. The unit is part of the UK Special Forces Group.

The principle role of the SFSG is to provide direct support to UKSF intervention operations, as well as reinforcing UKSF in other key capability areas such as provision of specialist training and support to domestic CT operations. It has a specific specialist infantry role and personnel are equipped and trained accordingly.

The Special Forces Support Group (SFSG) is Britain's newest special operations unit, formed around a core component of members of the 1st Battalion Parachute Regiment, Royal Marines, and some thirty members of the RAF.

An SBS diver coming ashore, fully armed and ready to fight.

SPECIAL BOAT SERVICE (SBS)

At the end of the 1940s, 1 Special Boat Section was formed by the Royal Marines Amphibious School. In 1950, 2 SBS was formed from a detachment of the Amphibious School attached to the Royal Navy Rhine Squadron in West Germany. The following year saw the formation of 3 SBS, with 4 and 5 SBS being formed from the Royal Marine Force Volunteer Reserve. In addition, a further section was formed from volunteers and attached to 41 Independent Commando RM for operations in Korea. Shortly afterward, 6 SBS was formed in Malta for operations in the Mediterranean, subsequently accompanying 3 Commando Brigade during the landings at Suez in November 1956.

During the mid-1960s, 1 and 2 SBS were deployed to the Far East on operations during the Borneo Confrontation, patrolling and establishing observation posts on the rivers and coastal areas of Sabah and Sarawak. Meanwhile, 6 SBS was tasked with carrying out beach reconnaissance throughout the Arabian Gulf, being joined there by a detachment of 2 SBS.

During the early 1970s, the Special Boat Company (redesignated the Special Boat Squadron in 1975) became involved in anti-terrorist operations. In 1971, SBS personnel were deployed to Northern Ireland, initially as part of the Military Reaction Force (MRF), but subsequently with 14 Intelligence Company. The year 1971 also saw two members of the SBS, together with an EOD officer and a 22 SAS NCO, parachute into the Atlantic in response to a bomb threat aboard the liner QE2. In April of 1973, an SBS protection team was on board the QE2 during a cruise carrying Jews from America and Europe to Israel.

In 1975, the Royal Marines were tasked with providing Britain's Maritime Counter-Terrorist Force and 1 SBS was thereafter dedicated to this role. In 1979, 5 SBS was deployed to Arbroath, Scotland, in support of Comacchio Company, the quick-reaction force for the protection of nuclear weapon sites and convoys and for counter-terrorist operations on ships and offshore installations. From there, 1 SBS concentrated on operations involving ships. In 1987, 1 and 5 SBS were amalgamated and located at Poole in Dorset where they were redesignated M Squadron SBS and remained dedicated to the maritime counter-terrorist role.

In 1982, 2, 3, and 6 SBS were deployed at the start of the Falkland War, landing on the islands some three weeks before the arrival of the task force. SBS reconnaissance teams played a major role during the campaign; keeping enemy positions under constant watch and carrying out close and beach reconnaissance tasks.

In 1991, elements of the Special Boat Service (redesignated as such post-1983) were deployed to Saudi Arabia during the Gulf War, carrying out Scud-hunting operations in the eastern sector of southern Iraq and attacking a major enemy line of communication. Two years later, SBS personnel were deployed to Bosnia on reconnaissance tasks and remained there until 1996 as part of the British Special Forces element serving with the NATO Allied Rapid Reaction Corps.

Today, the Special Boat Service, together with the three SAS regiments and two SAS signals squadrons, forms one of the elements under the command of the Directorate of Special Forces which was established in the early 1990s.

SPECIAL FORCES FLIGHTS/ SQUADRONS: RAF, AAC & RN

The SAS has had its own AAC flight since the early sixties. The helicopters are mainly used for surveillance, reconnaissance, and VIP transportation. In the event of a terrorist incident, the SAS will dispatch the Commanding Officer, along with the team commander, in order to assess the situation firsthand.

There are two Special Forces flights within the RAF, forming part of 7 and 47 Squadrons respectively and both dedicated to support of Army and Royal Marine special operations units. Aircrew within both flights are trained for special operations, namely low-level deep penetration into enemy airspace and the delivery and extraction of special forces.

The Special Forces Flight of 7 Squadron is based at RAF Odiham in Hampshire. It is currently equipped with the UK-designated HC.2 variant of the Boeing Vertol Chinook twin rotor heavy lift helicopter which can be fitted with four (two forward and two aft) M134 pintle-mounted mini-guns. Within the next two years, however, the flight will be re-equipped with the HC.3, which will be the first special operations-dedicated aircraft to be purchased by the British armed forces. The same aircraft currently operated as the MH-47E by the US Army's 160th Aviation Battalion—better known as Task Force 160—features a glass cockpit, terrain following FLIR, built-in fast roping brackets, and four .50 caliber Gecal mini-guns.

The Special Forces Flight of 47 Squadron is located at RAF Lyneham in Wiltshire. It is equipped with the C-130K variant (designated C.3 in the United Kingdom) of the Hercules transport fitted with an in-flight refueling probe, electronic counter-measure systems, and chaff and flare dispensers, both the latter designed to provide a measure of protection against enemy air defense systems. The aircraft are currently being equipped with dedicated night vision goggles (NVG) and compatible cockpit lighting. It is reported that four of the C-130J Hercules on order for the RAF will be for dedicated Special Forces use, being upgraded to the same specification as the MC-130E Combat Shadow operated by the special operations squadrons of the US Air Force. This aircraft has terrain-following radar and forward-looking infrared (FLIR) systems, an integrated avionics package for long-range, low-level covert ops enabling precision insertion, and resupply of special forces and NVG compatible cockpit lighting.

Finally, M Flight, of the Fleet Air Arm's 848 Squadron, equipped with the Commando Mk.4 variant of the Westland Sea King helicopter, provides support for the maritime counter-terrorist role.

SAS preparing to move into a rural OP.

SKILLS & TECHNIQUES

One of the main skills used by the SAS is surveillance and gathering. Surveillance is a technique used to obtain information, make connections, produce new leads, collate, and provide verification. The main usage of surveillance is military intelligence. Governments have long since learned that information gathered on the potential lethalness and capabilities of another nation help prepare for defense or attack. One major problem with military intelligence is the amount of information collected.

There are many forms of surveillance, and for the most part people believe that they can go about their everyday lives expecting a certain amount of privacy—this is not true. For the most part, every person in Europe is under some form of surveillance on a daily basis. They will not deliberately be observed or overheard, but records of their movements and actions will be recorded. Close Circuit Television (CCTV) cameras currently monitor city centers, major stores, gas stations, and motorways. Credit card transactions can be traced. Emails can be read, as can every stroke of the keyboard.

In the SAS, surveillance is carried out to identify persons who have been indicated in subversive actions. Human, electronic, and aerial surveillance techniques help establish a person's location, and behavior patterns leading to possible association with other personnel. Depending on the requirement, the surveillance can be a combination of static surveillance, foot, and mobile, with some technical devices thrown in for good measure. The latter is becoming more important in surveillance as technical devices become ever more sophisticated.

Most surveillance starts with capability and intent. Capability: I know how to make a bomb from just about anything in the kitchen, but I have no intention of doing so because it's dangerous and illegal (I have the capability but not the intention). Intention: I don't know how to make a bomb but as soon as I am able I will kill my neighbor. (I do not have the ability but I have the intention to commit harm.) Most SAS surveillance operations start by gathering information about people who have both the capability and the intention. However, as with all things, there needs to be a starting point.

EXAMPLE

A Special Branch officer would have several informers who would pass information either out of patriotism, fear, or for money. The information received would come from a wide variety of sources in Northern Ireland: farmers, priests, or even IRA members who had been coerced by Special Branch. A farmer may report he had seen a wire across one of his fields and the officer would pass this to TCG. TCG would task the SAS to investigate, who upon their finding the wire would trace both ends. A target reconnaissance would normally result in finding a bomb at one end and a trigger point at the other. The bomb was normally at a culvert under a road which would be packed with explosives. At this stage the Ammunitions Technical Office (ATO) would be summoned to disconnect the detonator while making sure the electrical circuit was still intact should it be tested by the IRA. An OP would be set up to observe the trigger point and wait for the IRA team to come and set off the bomb; once the bomb was safe, military and police traffic would be increased to entice the IRA into action.

CLOSE TARGET RECONNAISSANCE (CTR)

A CTR comes in many shapes and sizes. It may be a simple walk-past of a building or the actual penetration of the same property . The secret is to carry out your CTR without the knowledge of others. This all takes meticulous planning: firstly, what is your target, a house, factory, office, or even a park bench? Secondly how will you get to and from your target location? Thirdly, what is the aim when you get there? Finally, what equipment do you need in order to get the best results from your CTR?

The variations on the tasks presented to the SAS vary radically and there is no clear definition other than what is required for the individual operation. For example, if you are operating in London you might need a CTR to establish the best method of keeping an eye on a particular house. Is it best to try and locate a building opposite that provides good coverage of the front door? Or is it best to place a covert vehicle in the street?

Both static and mobile observation vehicles form an important part of any surveillance organization. They can be manned or unmanned depending on the requirements. As a rule of thumb, the larger the vehicle, the more suspicion it draws. For example, a large delivery van is more prominent than a normal car. A good walk-past of the location should reveal the type of vehicle best suited for the surveillance operation. Parking directly across the street is a bit obvious, added to which access to a parking space is not always guaranteed. Surveillance operators learn a lot of tricks when it comes to watching their target. Placing a video or fixed camera is easy, but it's making sure the device is not discovered which requires the skill.

Many terrorists, especially those living among a normal community, will have received some form of anti-surveillance training. If a strange vehicle turns up opposite their home for more than a few hours, they may become suspicious. One tried and tested method for those who think they are under surveillance is to approach the vehicle and bang on the doors, or if not obscured, look through every window. This has the effect of making the surveillance unit think they've been spotted and to replace the vehicle. Today we know this would be wrong. The sudden disappearance of the vehicle would only confirm the target's suspicions. A good surveillance unit would hold firm.

Camera systems have become digital, removing the need for photographic studios. Infrared film and covert flash adapters are widely used, allowing the CTR teams to make a visual record of their night's investigations without drawing attention to themselves. Long-range photographic lens are also used, allowing good quality photographs even at distances of over a thousand meters. Additionally, covert cameras can be placed in such a position as to trigger movement and so start surveillance. Almost every street in London is covered by a camera and people are being watched around the clock.

Author's Note: One of the best CTR operations I was involved with required that we should enter a certain house in a street in Belfast, Northern Ireland. The house was occupied most of the time, making entry into the premises difficult. Finally we entered the house by removing the back door of a bank on the same street. The door and its frame were removed while the alarm system was still operating. Once inside the bank we gained access to the roof and removed enough bricks which allowed us to make our way to the target house. A covert camera was installed with the power supply being taken from the lighting circuit. As a result of this surveillance, three men who were on their way to assassinate a district judge were arrested with their weapons.

The CTR requires images to be taken of the building indicating front, back, and side elevations, and a close up of the door lock is always helpful. The building material— brick, stone, or concrete—will also be noted, as will any entry or exit points. The position of external services, telephone, and electricity will also be noted and recorded, as will be the position of the garbage bin. All this initial information will aid any technical team which is required to enter the building and fit any audio/visual monitoring devices. Finally, several preliminary sites will be noted in order to set up a static OP which can best observe the target building.

GARBOLOGY

Garboloy is the name given to sifting through people's household garbage. It is a source of endless information about the household and the people living in it. It is a never ending surprise to note the amount of information people will happily discard thinking it will be incinerated or end up in a landfill. The location of the trash bin is a primary task for any CTR team, as is the time and days the bins are placed out for collection. Items recovered from a trash bin can reveal if people smoke and how much they smoke, while lipstick on a cigarette butt may reveal the presence of a woman. Bottles of alcohol or the lack of any may reveal the religious nature of the occupants. Bank statements and household bills will reveal a wide variety of information.

One of the most important things about a CTR is observation. For example, if there are no children's clothes on this line, it means only adults occupy the home. Yet the toy in the background would indicate differently.

CLOTHESLINE ASSESSMENT

When in position to observe a target's house with regards to collecting information, don't forget to look at the clothesline. Almost all households, including flats, hang some washing during the week, and almost certainly on the same day every week. Observation of the clothesline over a period of a month can provide the following information: number of people living at the address, their rough age, and their sex.

OP (OBSERVATION POST)

An OP can be set up anywhere and for any number of reasons. It may be to trigger the start of a surveillance operation or to provide "eye on," ready to take down terrorists about to plant a bomb. They can be situated in a roof of a house or deep in a forest; in some cases the SAS soldiers may well be buried or hidden in caves. The location is simply a means of attaining the best possible post for the task. The prime objective is concealment while taking advantage of archiving the best results.

OPs can be for a few hours to several days, and living conditions are generally very basic. No cooking, smoking, or noise. Long hours of inactivity and boredom backed up with cramped sleeping areas. These are all the pleasures that await an SAS manned OP, yet thousands have been done over the years.

Long-term observation in urban areas is best done from a static location, such as a house with a clear view of the target.

FOOT AND VEHICLE SURVEILLANCE TECHNIQUES

In general, targets are not followed once, but many times; in doing so the surveillance operators build up a pattern of the target's general behavior. In such a case, surveillance will be termed "loose," and the operators will remain at a safe distance to avoid being compromised. Surveillance on the target can be done in short stages until a number of known "triggers" can be identified, (i.e., at 5:05 p.m., they leave their place of employment Monday to Friday). Loose surveillance is normally carried out against a target that is living in a fixed location for a given period of time.

Both foot and mobile surveillance operations have three distinct phases: the trigger or pick-up, the follow, and the housing. Any operation will be based on the fact that you need a starting place, normally a location where you know the target to be or where they will be going. The surveillance aspect is to follow the identified target and finally place him in a known abode, the target house for example.

TRIGGER OR PICK-UP

Without the target being located there can be no follow, therefore it is vital to have a good trigger. The type of trigger used can be either static (best), mobile, or technical. If the target's house or place of employment is under constant observation by a static OP, then this can be used to provide the trigger. The commencement time for the operation will be deduced from static OPs events log. Surveillance units will be in position in good time and wait for the standby trigger from the static OP when they detect signs of target movement.

> **Author's Note:** Where the target is located in a difficult location such as a large car park, the trigger may well come from vehicle surveillance opera-tors covering either the target's vehicle or the exits. This is a difficult start and one which can easily lead to a lost contact. This can be avoided by knowledge of the target's route and the positioning of a second trigger vehi-cle at a critical location.

Both static and mobile triggers can be enhanced by the use of technical surveillance devices and tracking units attached to the target's vehicle. These can be covertly fitted to the target vehicle either on a temporary or permanent basis. Most modern devices are generally at rest when the vehicle has been stationary for more than fifteen minutes and are activated by the door being opened and the key being turned in the ignition. The technical tracking device sends a very accurate position signal on demand from the desk operator; this can be either constant or at predeter-mined timed intervals. The signal is displayed on a com-puter screen which has a street map overlay. The target's

As with vehicle surveillance, if you get too close you will be spotted. Close up in crowded areas and stay back when there are less people around.

vehicle position can also be moni-tored on a small mobile phone used by the foot or vehicle surveillance operators. Other technical triggers can be used, such as hidden micro-phones and cameras close to the tar-get's premises, but these are not as reliable as the tracking systems which continue to be of use during the follow.

Once the trigger has been acti-vated, the following phase progresses. In the case of a foot follow, the trigger will simply state the target is foxtrot, or with vehicle surveillance, the target is mobile. The surveillance will con-tinue until the team leader has decided that the target is housed. At which time the operation is called off.

RECORDING INFORMATION

It is not enough to simply follow the target around to see where they go; a detailed report must be kept. Foot surveillance provides the opportunity to see how much money the target is spending a month. How much do they spend in the supermarket, the pub, or on new items such a televisions and other electrical goods? The monthly total can be easily checked against the target's normal earnings. Several spies have been caught by just such a method, indicating they are receiving funds from an unknown source. Pointers to watch out for are as follows.

- Does the target lead a lavish lifestyle compared to known income?
- What type of credit card is the target using? Do they use the same cash point on a regular basis?
- Has the target got any sexual preferences? Do they visit gay bars or prostitutes, etc?
- How much alcohol does the target consume?
- Is the target compromising themselves (i.e., being seen with a rent boy)?
- Is the target a user of drugs?
- Places of frequent visits.
- Any unusual deviations in an otherwise normal route to and from a location.
- Does the target employ counter-surveillance tactics?

The reason for these questions and many more provide the surveillance team and others with vital information. If the target is the drug-taking type who likes to throw their money around; if married, do they have a lover? Visits to the same location on a regular basis may indicate a dead letter drop box. If information is found to be useful, it may be used to confront the target and turn them into a double agent. Every detail, no matter how small, must be recorded.

TRUE EXAMPLE

The IRA usually worked in cells which could contain around four members (men or women) who had known each other since birth, people they could trust. Around the edges of each cell would be like-thinking members who were used for less important tasks but if the opportunity arose, could be brought into a cell.

One such member became known to the Special Branch because he was having an affair with the wife of another IRA member who was serving a long prison sentence. If the IRA found out it would mean possible kneecapping (deliberately shot in the knee at close range, a standard IRA disciplinary punishment).

The SAS were tasked to take pictures of the man and woman in bed together and thus provide positive proof of the adultery. A few days later the fringe IRA member was stopped at a police vehicle checkpoint and shown the pictures. In exchange for not informing the IRA, and thus saving him a kneecapping, the man was asked to produce a very simple piece of information—so simple it could have been gleaned from any telephone book. The man readily agreed.

Several days later the man was stopped again and asked to pass the information while sitting in the passenger seat of a covert car provided by the SAS. In exchange for the information, the pictures and film were handed over—the SAS operator also produced an envelope containing £500 as a thank you. The IRA fringe man greedily took the money thinking how stupid the security services were. He did not realize that a camera hidden in the dashboard had recorded everything—he was now an informer and a dead man as far as the IRA was concerned.

Some weeks later two of the original IRA cell members were killed by the SAS and the informer was moved into the cell—the information on several shootings and bombings prevented many deaths and brought several senior IRA men to justice. Those can be considered as dirty tricks, but that was the nature of the war in Northern Ireland.

Mobile surveillance is difficult in heavy traffic.

MOBILE SURVEILLANCE

Mobile surveillance can involve any form of transport, from cars, vans, aircraft, and boats. The main attribute to mobile surveillance is the skill of the driver. The distance maintained between the surveillance vehicle and the target vehicle is dictated by the amount of traffic on the road and the environment. During rush hour on a busy street or highway, it is best to keep one or two cars behind the target; while out in the country it is best to stay out of sight if possible, just occasionally closing up to confirm you're still with the target after each side road.

Be aware of choke points such as roadwork or where the target can make unexpected moves or lock you in at traffic lights. Read the traffic in front of you and overtake any slow Sunday drivers. Road tolls are another source of holdups—if you can make it through the toll just before the target vehicle.

Make notes while driving so that you can record the target's average speed for the next folly. Additionally, know in advance where they are going if possible, as this will allow you to get one car in front. When using multiple cars, have some take faster routes and position themselves ready for any sudden turnoffs by the target vehicle.

Using a "Cat's Eye" marker to indicate where the pickup vehicle should stop.

SURVEILLANCE VEHICLES

Covert cars are used by the SAS for both surveillance and assault. In Northern Ireland, the SAS had a large fleet of covert cars all fitted with radios that under normal circumstances would not be found. The radio is connected to inductors, which are hidden in the roof above the driver or in the seat headrests. The operator hears any incoming messages by wearing an earpiece. A small button hidden beneath the internal carpet is used for transmissions; these can be either hand or foot operated. Other refinements to

the cars are cut-off switches to deactivate the tail brake lights, tracing devices, and the installation of cameras both for external and internal filming

Any vehicle selected for surveillance work must be mechanically sound and suitable for the area in which the surveillance is to take place. The vehicles should be a soft, nondescript color such as grey, with no distinguishing marks, such as front-mirror hanging dice or rear-window bumper stickers. The surveillance vehicle pool should be large enough to rotate the vehicles on a regular basis. They should be regularly maintained and fitted for use in all weather conditions. The outer make and type of vehicle can also be deceptive. For example, the rear trunk may indicate a model with a 1300 cc engine, while in reality the car will have been modified to take a much larger capacity engine. Other modifications include a powerful battery and improved radiator system to avoid overheating in long traffic delays. Internal temperature control is a necessity, as the surveillance occupants may be sitting in their car for long durations in unfavorable weather conditions.

Technical surveillance equipment is evolving with the ever-increasing presence of drones.

TECHNICAL SURVEILLANCE EQUIPMENT

The SAS possess a wide range of technical surveillance equipment, most of which is used by the anti-terrorist teams or for other covert operations. This equipment is either supplied and operated by 18 Signal Regiment whose members are SAS trained (264 Signals SAS) or, if something new is required, it falls to the SAS Operational Capability Cell. The latter glean the globe for any device that might be useful to SAS operations and it is not uncommon to see them at many of the premium military exhibitions around the world. Technical surveillance equipment is endless and comes in many forms:

- Cameras of every shape, size, and function including IR and Thermal.
- Miniature Closed Circuit Video Systems.
- Flexible Endoscopes for looking under doors and such places.
- Electronic tracking devices for car and humans.
- Intonation devices so small they can be hidden in weapons and mobile phones.
- Applications that will record any conversation or read any email or text from a mobile phone.
- Drones so small they can pass off as a bird.

COMBINED INTELLIGENCE ACTIONS

The SAS does not work alone in this capacity, as surveillance and intelligence works best when it is shared. In the collective manner, resources from various agencies pool what they know about a target and as such benefit greatly, as it provides for a more successful operation. The one great thing about British Intelligence is that it's shared with those who need to know, and those who don't . . . well they just don't know. The SAS is primarily a military unit within the British Army. In addition to its normal work, it is also tasked to do intelligence gathering and support other agencies such as SIS, Special Branch, MI5, and MI6.

SUMMARY

This chapter requires little in the way of a summary, as most of the text above is self-explanatory. However, it does help to explain the role of the modern day SAS and how through diligence and daring they have managed to integrate themselves into Britain's security services.

The requirements of keeping our country secure rests with these agencies and the tasks before them are constantly changing. Gone is the cold war and the James Bond spy image; today the modern terrorist lives among us, plotting and planning their horrendous acts. To combat this threat, the government must do all within its power to locate these perpetrators using every tool at their disposal.

While much general information is cleaned from the social median platforms and monitoring of mobile phones, the real work is on the ground. Locating the numerous terrorist cells and terminating them before they can strike. This is the hard end of terrorism but one the SAS fully understand and confront.

This chapter also highlights the fact that the SAS is just a small cog in the intelligence mechanism of Great Britain.

Chapter 9

THE GULF WARS

The invasion of Kuwait in August 1990 by neighboring Iraq created a situation which resulted in conflict with the combined forces of the United Nations. Although the UN attempted to negotiate with Iraq, it soon became apparent that military action was inevitable. Thus a huge Allied task force began to make its way to the Persian Gulf.

Battle plans were produced, dictating that the conflict would be fought along modern tactical lines. This meant sending in the Air Force to destroy Iraq's command and control centers and then once the enemy was blind, attack with massive and overwhelming ground forces. Although the British SAS had been on standby for some time, there seemed little scope for its deployment within this scenario. US Special Forces and US Navy SEALs had already taken all the border reconnaissance roles. It had been proposed that the SAS and Delta Force could be used to rescue the hundreds of foreign nationals being held as hostages in Saddam Hussein's "human shield" policy. However, this was a logistical impossibility, as the hostages were held in various locations all over the country. The risk was too high and the commanders considered that too many lives, both soldiers and hostages, would be lost, and this would lead to bad press back home.

Fortunately, the Commander of the British Troops, Lieutenant-General Peter de la Billiere, himself an SAS veteran and former commander of 22 SAS, managed to convince the overall UN Commander General Norman Schwarzkopf that the very special skills of the SAS would be invaluable. So in January 1990, the SAS left for the Gulf. It was the largest deployment of the Regiment since 1945, with men from A, B, and D Squadrons being present. Initial training took place at a secret location code-named "Victor" in the United Arab Emirates, prior to the SAS being sent to its forward base at Al Jouf in Saudi Arabia, from where it would penetrate the Iraq lines.

General de la Billiere decided that the SAS would prove most effective in creating diversions ahead of the main attack, destroying Iraqi communications facilities and tracking down the mobile Scud missile launchers, which so far had eluded both satellite reconnaissance and the air strikes. The SAS was to undertake these operations on the night of the January 22–23, six days prior to the anticipated start of the main hostilities.

In the event, the speed of activities took most people by surprise. Just before dawn on January 17, eight Apache helicopters from the US Army's 101st Airborne Division destroyed Iraqi air defense radars, creating safe corridors along which Allied aircraft could fly. The Air War had begun. Iraq, completely unprepared for such an attack, suffered substantial damage to its infrastructure and a devastating blow to its morale. In a desperate move to retaliate, Saddam Hussein turned his Scuds on Israel and launched twelve missiles on the suburbs of Tel Aviv. Miraculously, most inhabitants avoided injury, but the consequences could have been devastating.

As missiles, Scuds are not particularly accurate, but they do have the facility to carry warheads with a chemical or biological weapon capability. It was believed that Saddam Hussein had both these types of weapons of mass destruction, as he had already proven during the war with Iran and against the Kurds. Luckily the missiles that targeted Tel Aviv were carrying conventional warheads, but the realization of what could have happened caused mass panic. Israel threatened to invade Iraq and destroy the Scud sites and launchers. It also decreed that it would respond with a nuclear strike on Baghdad if Iraq used a chemical or biological weapon against their people. For United Nations Commanders this posed a nightmare situation. If Israel became involved in the war, it would cause massive disagreements among the Coalition's Arab allies and the whole war plan would be destroyed— something which Saddam Hussein was acutely aware of and the very reason for his Scud attacks. For the SAS this proved to be just the opportunity they were looking for and were immediately ordered to find and destroy the mobile Scud launchers. First the Scud launchers needed to be located and then attacked. The plan was simple—go in deep behind the enemy lines, observe the main supply routes (MSR), and destroy as many Scuds as they could find.

SAS in Iraq 1991, there were so many sergeants and warrant officers present, they actually held a mess meeting to discuss new furniture for the mess back in Hereford.

FIGHTING COLUMN

The bulk of the SAS were formed into half Squadron mobile strike units known as Fighting Columns, while some were formed into smaller foot patrols. The latter were road watch patrols and were all given the call sign "Bravo." The Bravo patrols were to be inserted by helicopter. These consisted of eight-man teams whose job it was to set up static OPs (Observation Positions) and survey the MSRs for movement of Scud launchers. Once the enemy was spotted, the hidden patrols would then call for an air strike to destroy the Iraqi vehicles.

The fighting vehicles in each column consisted of eight Land Rover-type 110, each of which was armed with a wide variety of heavy weapons. In each column, a Mercedes Unimog was used as the mother vehicle. This would carry the bulk of the extra stores, rations, fuel, ammunition for several types of guns, NBC (Nuclear, Biological, and Chemical) equipment, and spares. Motorbikes were used as outriders to scout ahead of the column. In this event both the fighting columns and Bravo patrols infiltrated deep into Iraqi territory, taking part in many separate search and destroy missions. The firepower carried by each column was enough to ensure that they could destroy just about every target or Iraqi unit they came across. Such mayhem would oblige the Iraqis to deploy large forces in order to locate the SAS.

SAS Fighting Column ready to enter Iraq and patrol deep behind the lines.

As the SAS almost always operates behind enemy lines, it is reasonable to assume that they carry some form of blood chit and currency with which they can buy their freedom or seek help should they be compromised and forced to run to evade the enemy, as happened with the ill-fated patrol Bravo Two Zero.

Author's Note: It is a standing joke within the SAS that much of the blood money gets

Author's Note: It is a standing joke within the SAS that much of the blood money gets lost in action. After the Gulf conflict was over, one enterprising soldier took his sovereigns to the local marketplace and enlisted the help of a local artisan. With loving care the goldsmith produced a solid gold winged dagger. The emblem is said to be cast from thirteen half sovereigns of the issued blood money, two more of which were given to the Arab jeweler. The five surviving coins remain intact. This has caused some dilemmas, as the Gold Winged Dagger of the SAS is of great interest to collectors, and there is no doubt that it would fetch a high price at auction, while on the other hand it is really the property of the Crown. The whereabouts of the Gold Winged Dagger remains a mystery to this day.

OPERATIONAL SKILLS

While the Bravo foot patrol had a distinct task of watching the MSRs and reporting back to HQ the moment for a possible air strike, the Fighting Columns were roaming about the desert solely for the reason of looking to pick a fight. These columns operated exclusively at night, preferring to lay-up during the day to avoid detection. All the SAS columns stayed inside Iraq for the full duration of the war. There are several good stories that highlight how effective both the Bravo and Fighting Columns performed.

One Fighting Column decided to stop just as it was getting light but by 6:30 a.m. they were still having difficulties finding a suitable location to hide during the daylight hours. Finally, with most of the vehicles camouflaged, the men started to prepare breakfast before going to sleep. At this point they heard a vehicle approaching; they had been spotted. An Iraqi vehicle was heading directly towards their position. One of the soldiers present told me this story, taken from the book *First Kill*.

SAS Gold Dagger.

Briefly looking around, I could see Pete getting himself behind the GPMG on the pinkie, while Tim, who was next to me, took up a fire position under the camouflage net. Del was crouched to my right and Mick was just behind me. At times like this, firepower is what it's all about. Amid the rush to get into position, none of us had even noticed that Mark was still sound asleep in his green maggot (sleeping bag).

We knew our location was not what it should be, and we had moved in, in a hurry. The Iraqis must have spotted us from some distance as they drove directly into the camp area and stopped within twenty meters of us. Luckily, they obviously thought that we were friendly forces, not thinking that enemy troops would be so close to Baghdad.

Mick whispered from behind, "What the fuck are they doing?"

I ignored the question and watched both the driver and another guy, who had been sitting in the front seat, get out. The driver then walked to the front of his vehicle, opening up the bonnet.

"Mick. Yorkie. Get ready. Cover me," said Del, "I'm going out to meet him." Del was about to play the hero. Breaking cover, he walked out to meet the Iraqi officer. It was incredible to see the enemy so close. The officer was a short, slightly chubby guy in his early to mid-thirties, tidy and clean shaven. He had dark hair and a very small moustache, characteristic of the Middle Eastern male. Del emerged from under the camouflage net, his weapon held to his side, away from the Iraqi's vision. The Iraqi approached; the look of bewilderment already written on his face. I could tell by his uniform that he was an officer, dressed in olive drab military trousers, shirt, and jumper. His head dress, a blue beret, had the Iraqi eagle emblem badge on it, and he was wearing the rank of a captain. What we didn't know was that the vehicle, a Russian Gaz 69, had more Iraqis in the back. The Iraqi officer, complete with map case and charts, walked towards Del. At a range of three meters, the officer realized we were the enemy, to which Del swung his rifle up and fired. Nothing happened. Automatically he dropped to his knees clearing the weapon stoppage—by doing so he had cleared my line of sight to the Iraqi. "'Breakfast in Valhalla you bastard." I fired.

The man fell dead. With that the world erupted with its lunatic song of battle, as everyone else opened fire. My eyes took mental stock of the situation, registering every little detail. I saw the driver's body do a little death dance before it finally crumpled to the ground. In a rush we all broke cover together, heading towards the vehicle. It never dawned on me at the time, but we had just initiated the first contact of the ground war, and that the first kill of Operation Storm, was down to me.

Following Dell and Mick, I ran around the front of the vehicle to check the back. The first thing I saw was Del pulling out a man who was very seriously injured. As Del dropped him from the vehicle, an incredible jet of blood gushed from his side. The wounded Iraqi now lay face down in the dirt with his knees tucked under his body and his backside sticking in the air. New noises distracted me from the grotesque sight, and I turned to see Steve tangling with a surviving, half crazed Iraqi soldier. Somehow this guy had cheated death; the rear cab where they had been sheltering was riddled with bullet holes. Steve was screaming at the Iraqi, who in turn wailed back in Arabic *himsh-ala* (praise be to God). Next he was being rushed off to the OC's wagon to be interrogated by two of our interpreters.

It did not take long to extract the news that a large force of Iraqis was on its way towards our location. This startling snippet of information had been obtained from the POW, and the two linguists had no reason to disbelieve him. At one stage of the interrogation, they thought they would have to shoot him to stop him talking. He just sang like a bird, pouring out numbers and locations nonstop.

"How strong is your unit?"

"About thirty thousand men." (translated from Arabic)

"Where are they now?"

"There. Just over there." (translated from Arabic)

"Holly fuck. Let's get the hell out of here!"

We all started packing up the wagons, very quickly throwing all the cam nets into the back, then piling on our bergens and ammo boxes. Once we were kitted-up and prepared for an immediate exit, two further problems came to light. The Iraqi dead were thrown into the back of their own wagon, which one of our guys would drive. The surviving prisoner was tied up and dumped in the Unimog. I jumped in my wagon and made ready to go. We drove off heading for the Saudi Arabian boarder putting as much room between us and the Iraqi army as we could. Eventually we stopped and called in a helicopter to extract the prisoner and the maps we had captured. Finally, we buried the bodies and placed a bar mine under the GAZ 69 and destroyed it.

Russian GAZ 69 captured by an SAS Fighting Column early in the war.

On the evening of January 22, 1991, eight members of a patrol using the call sign Bravo Two Zero were infiltrated into Iraq by Chinook helicopters. Their task was to observe the MSR and to sever underground communications cables which ran between Baghdad and Jordan; in addition they were to seek and destroy any Scud missiles in their area. Each member of the patrol was overloaded with huge bergens and additional stores needed to sustain them during their time in Iraq. Once the helicopter had left them the patrol moved some 20 kilometers, at which point they found a small cave in which to hide. The cave was about five meters high and cut into rock with an overhang which would conceal them from view and provide cover from fire. The patrol commander Sergeant Andy McNabb (real name withheld) soon realized that the radio they had was not working; this meant returning to the landing site the following night to RV with a helicopter and exchange radios. The patrol soon found themselves to be in a difficult position and decided to move, during which time they made contact with the enemy. A ferocious firefight

ensued and the patrol was forced to withdraw, heading for the Syrian border some 120 kilometers to the west. The journey was hard and dangerous, added to which the area was experiencing the worst weather in its history.

Through hypothermia and injury, the patrol became separated as a result of which four were captured, three died, and one managed to escape. Those that were captured by the Iraqis faced weeks of beatings and horrendous torture. In the end they were released and returned to the Regiment.

CONSIGLIO, TROOPER "BOB" ROBERT

There have been many books about the SAS patrol Bravo Two Zero in which an SAS foot patrol made contact with the enemy deep inside Iraq. They encountered freezing cold weather, enemy road blocks, disorientation, and endured the impossible as they fought their way out and ran for safety some hundreds of miles away. Two died of hypothermia, one was shot while engaging Iraqi soldiers at a road block, and one managed to make it to Syria and freedom. The other three were captured and tortured before being released at the end of the war.

A former Royal Marine, Bob Consiglio, joined the SAS in February 1990 and was posted to B Squadron's Mobility Troop. A year later, he was serving as a member of the ill-fated patrol Bravo Two Zero in the Gulf War. The patrol, operating on foot, was spotted, and the men found themselves running for the Syrian border, hotly pursued by the Iraqis. Just a few miles short of the border, Bob Consiglio was fatally wounded, yet he continued to give covering fire while the rest continued their escape. Trooper Consiglio was the first SAS soldier to be killed in the Gulf War.

SAS dress during the Iraq War consisted of anything that would keep you warm.

VICTOR TWO

The assault on the Iraqi installation known as Victor Two was given to one of A Squadron's half fighting columns. They received their orders as they lay hidden in the vast wilderness of the Iraqi desert. It began when the two signalers started to decode long incoming radio messages from SAS HQ in Saudi Arabia. This feverish activity around the two signalers quickly spread to the rest of the men and soon the orders to launch an attack on an Iraqi microwave station that night became clear.

The men began to prepare for the assault, with Mountain Troop putting together a model of the radar station from whatever kit they had lying around. They built it according to the intelligence coming over the radio which seemed quite detailed on the setup of the station but were confused where enemy strengths and dispositions were concerned.

At 3:45 p.m., everyone except for the duty sentries made their way to the middle of the LUP for a briefing. The sentries were to be briefed on their roles later; not an ideal situation, as in the rush, it was more than likely that they wouldn't get the whole story and would go into battle not completely sure of what was happening. Nevertheless, the column was behind enemy lines and sentries were essential.

The briefing started with the men being formed up in groups—the groups that they would be in for the assault. Once they had settled, Brian, the RSM, and Paul, the OC designate, joined them and the briefing began. Although Paul was a major and the new OC for the column, it was obvious to the men that Brian was still very much in charge (SAS rank structure has little to do with who's in charge).

The target radar installation was to be known by its call sign of Victor Two. The plan was for the column to drive to within 1,500 meters of it, place the vehicles in a fire support position, and then send out a recce team. Once the location had been confirmed, the team would return to their positions with the assault team and close fire support before guiding them to the target. The aim was to completely destroy the facility which was the main center for guiding the mobile Scud launchers to their targets. It had previously been attacked by stealth bombers but the problem was that most of it was underground and therefore safe from aerial attack. The only way to make sure it was taken out was by an attack from the ground. Once the column had achieved this objective, all vehicles were to make their way back to the present location.

Brian made himself the overall commander of the operation. There were to be three assault groups—each of three men, plus a close fire support and fire support groups. The rest of the men were to assist with driving the vehicles.

The objective itself was a building complex with a microwave tower of about 65 meters in height. The outside perimeter wall was about five meters in height while a three-meter-high internal security fence posed another obstacle. The main gate was guarded by a sentry position.

The plan was to fire two anti-tank missiles at the sentry position and the gate. An explosive charge would then be placed on the internal fence. Once this had gone off, the three assault teams were to enter the building, each taking a floor, including the floors belowground. The floors needed to be cleared and then lay the explosive charges before the teams got themselves out as quickly as possible.

The men, clear about their roles, soon got to work. They were enthusiastic about the operation despite the development of some internal politics within the group. The only piece of information that wasn't clear was on how many Iraqis there were at the installation and how well-armed they were. The intelligence coming through was vague but suggested that there were more civilians than military personnel present and so there was to be no indiscriminate firing on the part of the assault forces. A man needed to be identified as a soldier before he could be shot at. Of course, this was far from ideal in a situation behind enemy lines in the middle of the war. However, orders were orders.

It was suggested at the time that the column was being sent in because the regiment wanted to "blood" its soldiers; that is, to give them some experience of war and be able to claim one of its squadrons led an attack in hostile territory. It was also rumored that headquarters knew exactly how many Iraqis were at the base and how well-armed they were; after all, they knew all the other details about the base. Sensing that maybe they were not being told everything, the troops started to feel uneasy.

After a final brew up, the men formed up in their vehicles and drove for a couple of hours until they reached an Iraqi highway (or MSR as they were known). Light was now fading. Having checked that it was all clear, the group moved off again. Before they got themselves onto the road, however, they needed to bridge a ditch so that the Land Rovers could cross over. This was done by the Mobility Troop placing two sand channels, supported by sandbags in the ditch, so that the wheels could roll across them without the rest of the vehicle getting grounded. Within seventeen minutes, all the vehicles were over, the bridge dismantled, and they were on their way again.

Five kilometers (about three miles) from the target and the adrenaline started to flow. All senses were alert, expecting an enemy ambush at any minute; so far it had all seemed too easy. Then the ground started to become more difficult to drive on, causing the column to make a kilometer detour to the west. Here the ground became a mass of man-made slit trenches, but as the orders were to slow down or stop only if an enemy vehicle was spotted, the vehicles carried on at the same speed.

Closing fast now on Victor Two, the MIRA (Milan Infra-Red Attachment) wagon came forward, needing to continually observe the area for enemy activity. Now just 1,500 meters away from the installation, the column came to a halt as planned. One of the men left his bike at the location and it was logged in on the SATNAV so that the rest of them would have a position to make their way back to after the operation was over. The target was checked once more through the MIRA. It turned out that the site was huge—lots of buildings, vehicles, and people; soldiers were positioned in both slit trenches and bunkers. There seemed to be little sign of the "civilians" mentioned by RHQ.

At this point, Brian started to change the plans that were made at the briefing. He decided that they should move forward immediately, without the services of the recce group. The column drove closer to Victor Two, internal politics changing the lineup as certain members of the group decided they wanted to take the lead, no matter what. Driving along a tarmac road, they soon found themselves right in the middle of the enemy position with hostile soldiers and vehicles all around. With the previous orders changed beyond any recognizable form, most of the men didn't have a clue as to what they were going to do next.

Without warning, the lead vehicles pulled over to the left side of the road and the others were obliged to form up behind them. They were now parked alongside a small escarpment running parallel to the road. At last, the recce party was sent out to do a CTR (close target reconnaissance) and discovered the magnitude of the site. In all it was about a kilometer square in size; its focus was the massive control building and microwave tower bristling with communication dishes. They returned quickly and a revised plan was put forward. Now it was mooted that the vehicles would split into two groups and give cover from the flanks. Meanwhile, the assault teams, together with a cover party, would go on ahead by foot to carry out the process of demolishing the tower with explosives.

Their forward progress went unchallenged. On reaching the complex 300 meters from the vehicles, the demolitions teams went forward to lay their charges while the covering team hung back, keeping a lookout for any trouble. One of the cover party, while waiting near the back of an Iraqi truck, heard a sound from the cab. He went around to investigate and opened the cab door to reveal the sleepy face of an Iraqi soldier. The soldier, surprised, reached for his gun, despite efforts to stop him doing so. Then there was no other choice—within seconds he had been shot dead with a burst from an automatic rifle.

A firefight instantly erupted and the column felt that every gun in the compound was trained on them. Even the Russian-made anti-aircraft guns were turned on their position. The squadron fought back, giving it everything they had. By now the demolitions teams had placed their explosives in the tower and were moving back to the rest of the column. The order was given to move out.

As the vehicles slowly regrouped and withdrew, they drew down a heavy rain of fire. Nevertheless they kept going, punching a hole through the opposition with their own heavy guns. After an hour of battling their way through enemy positions, they disappeared into the night. Remarkably, despite the dangerous internal political maneuverings and the change of plans, there were no casualties. A Squadron spent a total of forty-five days behind enemy lines, during which time two were killed and one was captured. The squadron received a DSM, two MCs, four MMs, and four MIDs for their bravery.

CTR IN IRAQ

The incident took place during the early hours of February 9, when a three-man A Squadron team were carrying out a CTR (close target recce) on a communications installation. The team was commanded by the SSM (Squadron Sergeant Major). The SSM was standing on the back of a Pinkie (SAS desert vehicle), observing through the MIRA. The team had already cut their way through the barbed wire which surrounded the installation, leaving one Pinkie at their entry point to act as cover. Moving further into the Iraqi position, the SSM realized that the place was far bigger than they had originally thought. Making a snap decision, they decided that their only alternative was to brass-it-out (be bold) and drive through the Iraqi position as quietly as possible, hoping that the dark would conceal their identity. Suddenly an Iraqi soldier stepped out in front of the wagon and made a vain attempt to stop them—he was swiftly dispatched. Seconds later, the Pinkie was taking incoming fire from all directions. The SSM was hit in the first burst. A round had gone through the top of the knee, throwing him into the back of the wagon. At the same time the driver shot forward, accelerating hard, hoping to burst through the position. He had just driven through the camp gates when their luck ran out. Three hundred meters from the camp, the wagon lurched to a sudden stop as it crashed hard into a tank ditch (it was pitch black and there had been no time to switch to PNG). Seconds before, the SSM had raised himself, hoping to get the rear gun into action. The sudden stop threw him over the roll bar, where he landed half on the laps of the passengers and driver and half on the trunk. His wounded leg was lying twisted on his chest and he sustained severe damage to his hip and back. Shaking themselves, they immediately made to escape dragging the SSM between them. They did not get far; the damage from the bullet and the vehicle crash had left him in critical condition. The other two SAS soldiers dragged the SSM up a slight rise, where the three hid among some rocks. Here they attempted to staunch the flow of blood that was coming from the SSM leg wound, but the injuries were extremely bad. Slipping in and out of consciousness, the SSM commanded that the other two should leave. One of the two asked if he should finish off the SSM (shoot him), as capture by the Iraqis would mean torture and possibly death. It was a tempting offer, but the SSM declined, willing to take his chances.

The men did as ordered, taking with them all the weapons and leaving the SSM with just a 66mm disposable rocket. They ran for several days heading for the Saudi Arabian border. Their distress beacons were pinged by American jets returning from a mission and relayed back to Coalition HQ, at which stage an SAS fighting column was vectored to intercept the runners—this they did successfully.

Alone, the SSM slipped unconscious once more, only to be woken in daylight by Iraqi soldiers who were looking around the wrecked vehicle some 300 meters away. In one final act of defiance, the SSM operated the 66mm rocket and prepared to fire it into the middle of the soldiers. At the second he was to fire, a pistol was placed at his head as a second group of Iraqis found him, and he was taken prisoner.

Surprisingly, the SSM was treated fairly by an Iraqi civilian doctor, who did a wonderful job of repairing the damage to his injured leg. The doctor had done his medical training at Bradford in England and had many fond memories of the country. Eventually the SSM was treated in a hospital, where his interrogation continued. One day, the two interrogators were standing by his bed deliberating a question to ask him; as they turned to speak he preempted the question, giving away the fact that he spoke fluent Arabic. Despite protests from the medical staff the two interrogators ripped the drips and such from his body and gave him a beating. Nevertheless he survived and was finally repatriated via the Red Cross at the end of the war.

SAS deployment in the Second Iraq War, this time was house-to-house searches for the Iraq leadership.

SECOND IRAQ WAR

Author's Note: Since 2003, the amount of public awareness of SAS activities has slowly come to a standstill. All media is avoided by its members and any breach is followed by an immediate dismissal and RTU. Even senior officers who have made the odd or slightest remark about SAS operations have been severely disciplined. Most stories now emerge as a result of a very close connection with a member of the SAS and several drinks; even then it's difficult to get a detailed story.

What has been clearly established is the amount of work the SAS Squadron has undertaken while operating in Iraq and Afghanistan. There are those members of the regiment that find the situation and work in Afghanistan thrilling, but for the majority it has become a real grind and a very dangerous one. In days gone past, Squadrons returning from a tour in the Middle East could be recognized by their healthy suntanned appearance; those returning from Afghanistan in 2009 are recognized by their gaunt, transparent skin. Continuous night operations, confronting the enemy almost every day, have taken a severe toll on the regiment.

For the SAS, the second war against Iraq started out in a similar way as the first one: preparation, rehearsals, and insertion. From the outset, however, there were differences: while one Squadron would carry on as before and patrol the area north of Baghdad, the others were given more specific targets. One of these included painting the main Iraqi bridges in and around Baghdad with a laser; equally communications hubs and other government buildings received the same fate. In fairness this provides aircraft with a much greater degree of accuracy. While this tactic was vital, getting into Baghdad before the

Uday's Palace, home to the SAS and US Navy SEALs.

Americans arrived proved a very difficult job. Once the ground war was over, there was a slight lull; most of the Iraqi army and police had simply surrendered or left their posts and gone home. But the peace did not last. The initial bewilderment gave way to anger, and because of the chaos, past feuds between various religious groups had time to surface.

To understand the complexities of operating in Iraq, one must first look at the diverse problems. First and foremost there was a lot of tension between the various factions in Iraq, factions which used the presence of the coalition forces as an excuse to settle old scores. Secondly, there was the problem of finding the enemy. But most of all there was the issue of internal security. The public does not understand the amount of pressure that is put on the SAS Squadrons operating in Iraq. On the one hand, there will be pressure from the British Government for them to show decisive action and bring home a good result, added to which is the pressure from the American Administration who want the same thing but wrapped up in a different way. Political pressure will often override SAS operations. For example, if they discover proof that Iranian agents are involved, the British Government may well wish to avert any clash with Iran should this be brought to public knowledge.

However, the worst pressure comes from intelligence. It is almost impossible to keep intelligence a secret. When the SAS finds something, it has to be shared in order to gain advantage from it. For example, when the SAS carry out a raid, one of the items which produces a large amount of intelligence is the mobile phone. These can be scanned for names, contacts, conversations, and messages. This, however, is a specialist skill and at the time only available to the Americans. While American security is extremely good, at least a dozen people will have access to how the mobile was obtained and what intelligence was retrieved from it. These in turn have masters they must confide in and so the circle grows, and the larger it gets the bigger the security gaps become. At other times the spooks will forestall any immediate action against a suspect because they have someone close to him (an Iraqi spy working for the British or Americans) and do not wish to blow their cover. This strange situation happens within the SAS all the time. It evolved out

General McChrystal, an excellent commander much liked by the SAS.

of Northern Ireland, where IRA suspects were allowed to go unchecked just to protect the source—for the SAS this is just one more of the complexities that make life very frustrating. Luckily for the SAS, the man in charge of Special Forces Operations in Iraq was General McChrystal, commander of JSOC.

Author's Note: The Joint Special Operations Command (JSOC) is part of the American Special Operations Command—referred to as SOCOM. JSOC are in charge of special operations. It was originally established in 1980 on the recommendation of Colonel Charlie Beckwith in the aftermath of the failed attempt to rescue hostages in Iran (Operation Eagle Claw). It is located in Poe Air Force Base and Fort Bragg, in North Carolina (home of Delta Force), with command elements in Tampa, Florida.

In the United Kingdom, Northwood Headquarters is the home of the Permanent Joint Headquarters (PJHQ). Northwood is a modern command, control, and communications facility installed to allow both the Chief of Joint Operations to direct national operations worldwide and the Commander in Chief to direct Maritime Operations; Northwood is the hub from which all British wars and conflicts are controlled.

It was decided to form PJHQ at Northwood in July 1994, as a part of the Defense Costs Study, to replace the previous approach whereby headquarters' staffs were drawn together ad hoc in response to a developing crisis, as had been the case during the first Iraqi conflict. The primary role of the PJHQ, therefore, is to be responsible for the planning and execution of UK-led joint, potentially joint, combined and multi-national operations. PJHQ also runs Operational Bases around the world: PJOB Cyprus, PJOB Gibraltar, PJOB South Atlantic, and PJOB Diego Garcia covering the Indian Ocean territories.

General Stanley McChrystal led the JSOC between 2003 and 2009. From 2005, after high casualties had been sustained within the American Special Forces, General McChrystal formally asked the Director of UKSF if they could assist. Commanding officer of the SAS was Colonel Richard Williams who had had a distinguished Special Forces career. CO 22 SAS managed to build a relationship with McChrystal, and as a result, UK Special Forces, including the SAS, SBS, SRR, and SFSG in Iraq came under JSOC. Not everyone agreed with this strategy and there was some dissent between the OC 22 SAS and DSF. This was eventually resolved, and resulted in the SAS remaining in Iraq while the SBS were sent to Afghanistan.

One of the first SAS operations in Iraq was Operation Row, which started in March 2003. The intention of this operation was to deceive the Iraqi High Command into thinking that an invasion would also come from the West and North of the country and keep Iraqi divisions in those areas from reinforcing the Coalition's main invasion route.

In Operation Row, the regular 22 SAS Regiment worked alongside the Australian SAS Regiment and American Special Forces. The UK task force

Australian and British SAS capture H1, H2, and H3 airbases in March 2003.

involved the deployment of the CO SAS, RSM, Operations Officer, and a HQ Element. B Squadron SAS came in from the West in lightly armed Land Rovers, rather like the originals in the Western Desert hunting for targets which posed a threat to the Coalition troops such as Ballistic Missile Launchers. Meanwhile D Squadron seized a desert airfield and UK SBS's M Squadron came through the South. M Squadron faced a battering in a firefight and ended up losing equipment. Another SAS team was detailed to support UK 1 Armored Division and be infiltrated into Basra with SIS Officers. American Special Forces were responsible for coordinating Kurdish factions in the north.

The role of the SAS in capturing an Iraqi Missile Base and two strategic airfields has been well-publicized. All of the targets were heavily defended and the raids were all joint operations. The Special Forces teams were sent detailed satellite imagery straight to their Panasonic CF19 laptops, carried in their bergens, of the various locations that needed to be targeted. (All SAS teams carry a two-way satellite link which is supported by 18 Signals Regiment.) In addition, the teams were sent up-to-date detailed intelligence about the proposed targets which posed a threat to the main coalition effort. Until the moment when the order for the operations to "go ahead," the British SAS carried out detailed static and mobile surveillance of the targets and developed a detailed pattern of life of the airfield.

Under orders, the SAS carried numerous assaults on Iraqi forces using lightweight Land Rovers with .5 caliber machine guns and automatic grenade launchers. The assault teams arrived suddenly, overwhelmed the surprised Iraqi forces, and spread confusion. Every SF vehicle is self-sufficient, carrying a specialist team of medic, linguist, and patrol communications expert and vehicle mechanic. All the petrol and rations needed are carried by the team with emergency equipment in bergens in case of a serious firefight resulting in the loss of a vehicle. The tactics were also reminiscent of the Second World War, as numerous teams of armored vehicles, both British and Australian, attacked the airfields with speed and surprise. Outflanking teams provided covering fire as the main assault made directly for the airfield buildings before splitting up to storm the control tower, offices, watch towers, and hangars. Teams of four dismounted to clear the buildings, carefully marking each room as they did so. Less than twenty minutes after entering the airfield, most of it was secure, with prisoners being bundled to the rear for interrogation. Once the mission was complete, regular troops were brought in to secure the perimeter and establish control—ready to reuse the runway for more UK and US forces.

With the war well underway and with their initial tasks completed, most of the SAS made their way into Baghdad. The war still raged in the city with mortar and artillery shells falling as the SAS Chinook landed. The party included Brigadier Graeme Lamb, then Director of Special Forces, and members of SIS. The main effort for the UKSF in Baghdad was to help SIS establish a new station. At this point, Brigadier Lamb linked up with the CO of 22 SAS.

It was a credit to the Director of Special Forces, his operations staff, and the professionalism of the UK SF that not a single man was lost in Operation Row. Once the offices were established, SIS needed the SAS to assist the station in providing timely intelligence for three main subjects: the search for the legendary Weapons of Mass Destruction, general political intelligence in Iraq, and finally the all-important information forecasting what was going to happen next.

In May 2003, owing mainly to the severity of the initial conflicts, G Squadron replaced B and D Squadrons who were gradually returned to the United Kingdom. G Squadron had not been involved in Operation Row and had been in the United Kingdom in preparation for counter-terrorism operations. Once the SF HQ had been established in Baghdad, CO 22 SAS and his group also returned home to Hereford, leaving OC G Squadron to carry on. Many of G Squadron members had been trained to serve as native interpreters and so gained crucial information from captured Iraqis. This tactic proved to be highly successful as the war progressed. The intelligence gained from an immediate arrest was of a much higher value than that extracted once a prisoner had had time to reflect on his situation.

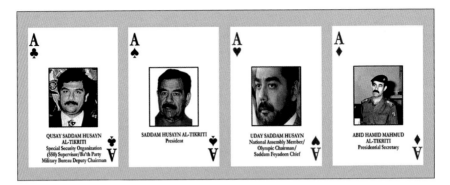

Deck of cards to identify the most wanted Iraqi personnel with the Hussein family at the top of the deck.

In June 2003 a new Operation, Paradoxical, started. The operation remit was for the UKSF to target threats to the coalition. By now the coalition had published the famous deck of cards of the most wanted Ba'athist party members including Saddam Hussein and his two sons. The first major test for G Squadron SAS came on June 16, 2003. One of those on G Squadron's list was Lt. Gen. Abid Hamid Mahmud al-Tikrit who, according to intelligence developed from a local source, was hiding out in Tikrit. An SAS troop dashed to Saddam Hussein's backyard in their lightweight Land Rovers and met a group of US SF troops. A rapid plan was devised and the Ba'athist party member was easily caught.

It is also believed that an SAS unit captured Watban Ibrahim Hasan al-Tikriti, Saddam Hussein's half-brother. He was apprehended close to the Syrian border where SAS units had been waiting in ambush. Watban, a presidential adviser, was accused of overseeing executions, deportations, and torture. However, he fell out of favor when he argued with Saddam's son. Uday; it is rumored that Uday shot Watban in the leg at a party back in 1995. The SAS handed him over to the Americans.

On June 24, 2003, a few months after the invasion, six British Military Policemen were butchered during a firefight in Majar al-Kabir. Two days later, SAS members were deployed to the town; the team made inquires as to what happened but soon came under fire. Their mission had not been helped by the nearby Parachute Regiment declining to get involved. The SAS group was forced to retire while the

British Divisional Commander later discouraged the SAS from going back to get those responsible. One SAS source later remarked that it would have been difficult to extract the ringleaders, but the spirit of the Regiment determined that G Squadron would want to try.

The code name for overall UKSF operations in Iraq was changed to Operation Crichton. It was now A Squadron's turn to be deployed and they replaced G Squadron who had completed a successful four months in October 2003. The arrival of A Squadron brought a renewed offensive spirit to Iraq. The SAS Squadron were now adopting a covert approach by wearing local clothes bought from the markets and driving locally owned civilian vehicles to fit into the new environment. The problem with these vehicles is that they did not have much capability. Luckily, RAF Chinook helicopters had by now arrived at Baghdad airport and A Squadron put them to good use.

Towards the end of the year the SAS had settled in. On October 31, 2003, A Squadron had orders to assault two houses in Ramadi as part of a covert operation code-named Abalone. The Mission was to capture a number of foreign Jihadists from the Sudan. The SAS and SBS assault teams were supported by American SF in Bradley Armored Vehicles. The first building was assaulted without any major issues but the second assaulted by SBS resulted in an immediate firefight with an RGP round exploding nearby. This was followed by masses of AK-47 rounds which hit all members of the assault team. One member of the SBS team, Corporal Ian Plank, was killed in the operation with another SF corporal seriously wounded. Without thought for their personal safety, two SAS troopers returned under fire to get their seriously wounded comrades. The Squadron regrouped and assaulted once more, but casualties and the stiff resistance curtailed the operation. While the objective of the mission had not been achieved, the SAS/SBS team had detained four non-Iraqi fighters in the night's operation. Corporal Plank of the SBS was the first Special Forces member to die in the second Iraq war. The SF then became embroiled in several missions in Fallujah with A Squadron raiding several prominent buildings in the city. The battles in Fallujah lasted throughout October and November 2003.

The dress of these soldiers indicates how interactive the SAS and American Special Forces were.

In February 2004, A Squadron was relieved by B Squadron, who kicked off with Operation Aston. The intelligence had been gathered by SIS which had been concentrating on how foreign fighters were traveling to Iraq. The operation was judged a success as they had managed to kill two foreign fighters. Moreover, it was discovered that the two KIA were members of a known terrorist group called Lashkar-e-Taiba. Some fighters were captured and later provided valuable intelligence for the coalition.

By mid-2004, the pace of operations seemed to be dropping; in 2003, A Squadron had completed eighty-five raids in a four-month tour. D Squadron returned to Iraq and prepared to take part in the November 8, 2004, assault on Fallujah, an area which still refused to be settled. Conversely, it was deemed that risking Special Forces in what was essentially seen as a FIBUA operation was a waste of manpower and consequently a battle group of the Black Watch were sent in to Fallujah.

Early 2005 saw a number of changes. The SAS gained a pair of Pumas and became semi-independent as far as travel was concerned. Helicopters were needed owing to the challenging environment of improvised explosive devices (IEDs) which had now come into their own, with one or several going off on an almost daily basis. Added to which, there was growing resentment towards the Coalition Forces and the insurgent numbers were growing. They had developed an impressive number of lookouts and sentries to counter surprise attacks by the Coalition Forces. To counter the IED threat, the SAS exchanged Land Rovers for armored vehicles, as it was becoming dangerous to drive lightweight vehicles with no armor.

On January 30, 2005, an RAF Special Forces Hercules aircraft, which had just taken off from Baghdad airport on its way to Balad, was shot down by small arms fire. Six minutes after takeoff the radio operator reported a fire on board and the aircraft was reported missing twenty-four minutes later. Following investigations, it was revealed that a projectile had penetrated the starboard wing fuel tank, causing a fire in the wing. The ensuing explosion caused the aircraft to crash. Ten people died as a result, one of whom is believed to have been a member of 264 SAS Signals Squadron. G Squadron mounted an operation close to Baghdad airport to hunt down those responsible.

On July 23, 2005, JSOC had acquired urgent intelligence to the effect that multiple suicide bombings were about to take place. To forestall and intercept the bombers, Operation Marlborough was launched. The building which housed the bombers was identified and kept under surveillance by a US Predator UAV, which supplied real-time video of the building and the surrounding area. A task force of American Special Forces, together with M Squadron SBS and some troopers from G Squadron, prepared to assault the building. Eventually the plan changed and it was decided to wait until the bombers exited the building, for fear that one of them may detonate a device should they see the assault coming. Their patience was rewarded when, at around 8 a.m., three of the suicide bombers were shot by SBS snipers as they left to complete their deadly mission. At this juncture the building was stormed. One member of the SBS managed to shoot another suicide bomber before he was able to explode his bomb vest.

On September 19, 2005, two SAS soldiers who were returning from the Kuwait border had a brush with the Iraqi police. While the news reported the men had been on a surveillance operation, in truth they had been escorting two members of MI6 to the Kuwaiti border. On the return journey their car had broken down and so they hijacked a local car, transferred their equipment, and proceeded on to Basra.

At some stage they ran into an Iraqi police roadblock and for reasons known only to themselves decided to make a run for it. Shots were exchanged and the police gave chase. Unfortunately the car they had hijacked was in poor shape and they were forced to stop and surrender to the pursuing police. They were then taken to Jamiyat police station in Basra. However, they were able to send off an emergency message to the duty signaler in Basra Palace, warning of their imminent capture, but it is not known if he received it.

Given that the two SAS soldiers were dressed like local Arabs and their car was full of weapons including a rocket launcher—and given that at least one of the police had been shot during the chase—the Iraqi police were not very happy. Subsequently the men were tied up and badly beaten with the process being filmed.

Not sure if their message for help had gotten through to the duty signaler in Basra Palace, the SAS shouted for help when they heard English-speaking voices close by. They shouted out that they were Special Forces and being held prisoner. It is believed that the prison was being visited at the time by a British Officer who was training the police— either way the alarm was raised.

Two SAS men arrested while returning from escorting two MI6 operators to the Kuwaiti border.

The order to release the two SAS men was ignored by the Iraqis and when OC A Squadron met with Brigadier Lorimer, he found that a rescue attempt would not be sanctioned by PJHQ. By this stage all intelligence agencies, including the Americans, were focusing their attention on the police station. It is said that intelligence intercepted communications from the station indicating that the SAS men would soon be transferred to the Iraqi Hezbollah—this was not true.

Back in the United Kingdom, the MoD were being pressed for permission for a rescue attempt to be made before it was too late. However, the most senior figure, an RAF commander, was out playing golf and his deputy, Major General Peter Wall, was out of the office. It was at this stage that the SAS commander took it upon himself to order the rescue attempt.

This stage of the operation included tanks, armored personnel carriers, and helicopter gunships. The police station walls were breached and the two SAS men freed. With the station all but totally destroyed, most of the Iraqi prisoners being held also managed to escape during the mayhem.

By 2006, SAS operations in Iraq had reached a fever pitch. The men were expected to go out on operations every night; every operation was to be completed; every operation was to produce useful intelligence. This was the McChrystal approach to covert operations in Iraq. His belief was that insurgency could only be overwhelmed by a relentless tempo of operations. Joint intelligence gathering missions were the way forward.

This joint strategy highlighted several important issues, the main one being that the SAS always appeared as second-class citizens compared to the superbly equipped American Special Forces in Iraq. The SAS were forced to beg, borrow, and even steal from their American cousins in order to complete their missions successfully. It was not until May 2007 that the SAS in Iraq were provided with new vehicles and improved night vision aids.

The key strategic benefit of closer integration with JSOC was the increased flow of intelligence through mobile phone intercepts provided by NSA, as well as other sources of information which could now be processed by the American intelligence institution. However, it must be stressed that the SAS contribution to JSOC in providing significant intelligence material was enormous.

Author's Note: Most SAS operations took part during the hours of darkness as this provided good cover on approach and withdrawal. Additionally, there was always a curfew, which meant that anyone on the streets was fair game. Night vision goggles are extremely useful, but in truth these were extremely limited to the SAS and most members had to rely on the mark one eyeball. What really counts is stealth and surprise. Get into the enemy's area of fire before he knows you're there; that way you have the upper hand.

One must also understand that the SAS is extremely limited in its numbers; it always has been and most probably will remain that way. This simply means those that are in theatre work longer and harder. During the six-month rotation of 180 days, A Squadron carried out some 175 operations. Basic math tells you that every single night these men go into combat. This was not simply a matter of going out on patrol; these were combat assaults, targeting houses and kidnappers. Delivery to the site was by vehicle or helicopter to a point several kilometers from the target, after which they would walk in order to achieve some surprise. Combat like this, night after night, takes its toll even on SAS soldiers.

On March 9, 2008, during B Squadron's tour in Iraq, the intelligence team developed an operation involving the radical cleric Muqtada al-Sadr's Mahdi Arm. He was reported to be holed up in the Sunni triangle, and owing to pressures on the local American SF team, the SAS were asked to deploy at short notice. By March 26, B Squadron finally had enough intelligence to mount an assault. SAS teams approached the house while the SFSG were covertly establishing a cordon. It was standard operating procedure to send a dog into the building to check for the target. When the dog did not appear, owing to the seriousness of the operation, the SAS assault team deployed anyway. Four SAS men were wounded and Nick Brown was hit in the chest and fatally wounded in the firefight within the house. A general firefight broke throughout the neighborhood. The SAS disappeared, but the local American Army unit faced a series of ground battles with locals angry at what happened. Nick Brown died later of his wounds.

One of the many missions that fell to the SAS in October 2008 was the attempted rescue of British and American nationals who had been taken hostage in and around Basra. One such operation took place in a block of flats which the SAS assaulted. Two SAS soldiers entered the apartment by forcing open a barricaded kitchen door. As Sergeant Hollingsworth entered the kitchen, a shot was fired and he was hit in the chest. There was some discrepancy of opinion as to whether the 5.56mm bullet was fired by British troops or Iraqi insurgents hiding in the building. It was known that the terrorists used 5.56mm ammunition, which makes it almost impossible to say which side actually fired the fatal round, although all the rest of the SAS assault team claim not to have fired during the raid. No weapon cable of firing the 5.56mm round was found in the apartment. Iraqi women—along with children—were the only ones in the apartment and it was possible they could have concealed a weapon knowing they would not be searched. A number of suspects were detained, but it is known that two—one of whom had been seen on the balcony outside—managed to escape. Shortly after the shooting, the operation was aborted and Sergeant Hollingsworth was taken to a nearby military hospital where he was pronounced dead from a single gunshot to the chest.

SUMMARY

During the first Gulf War, the method of fighting almost reverted to that of the original SAS in World War II. Fighting columns penetrating deep behind enemy lines caused chaos and destruction in the enemies' rear. Surveillance of the MSRs by standing foot patrols was not as successful as the story of B2 Zero illustrates.

The second Iraq war was entirely different with most of the SAS tasked to locate the former Iraqi leadership. This meant house-to-house searches, which turned into shoot-outs on many an occasion.

DRESS & EQUIPMENT

The SAS have never been the best dressed military unit in the British Army; dress is individualized and more suited to the current climatic conditions, concealment, and at times protection. Hence dress for the first Iraq war was standard desert uniform covered with whatever thick, warm jacket that could be found. A Shamag (Arab headdress) was worn to protect the head, although most preferred to use it as a neck scarf. Gloves and mitts became a necessity as many soldiers suffered with acute skin cracks that soon became infected. Webbing consisted of water, survival kit, and lots of ammunition pouches. As they were deep behind the enemy lines, it was always advantageous to carry the British flag in order to identify themselves to friendly aircraft.

Author's Note: Most people believe the Middle East to be hot all the time; during the Iraq war temperatures fell well below freezing with high winds and snow showers. Initially the SAS were ill-equipped for the weather but soon overcame the problem.

WEAPONS

Individual weapons varied but were mainly the M16 and M4, some being fitted with M203. A few personnel also carried a Welrod silenced pistol, something the SAS had not used for years. Most of the SAS were assigned to the fighting columns and manned heavier weapons attached to the vehicles. These included the Browning .5 heavy machine gun, GPMGs, American Mark 19 40mm grenade launchers, and Milan anti-tank missiles. The latter were excellent for night operations; being fitted with thermal imaging sights it could be used up to 8km away, even in total darkness.

The SAS Squadrons operated at roughly 2/3 strength, the remaining personnel being assigned to various other tasks within SIS. In 2003, the SAS were based out of Oday, Hussein's palace, with the Navy SEALs next door. From 2004–2006, the squadron was based out of the old embassy and Iraqi government buildings during the Task Force Black days. Living conditions were OK and most men had real single beds which replaced the old cot beds and guaranteed the much needed rest between operations. Owing to the amount of collateral destruction, electricity was limited and drinking water was restricted to bottled water. On the plus side, there was access to US dining halls and PX facilities for the SAS.

In the early days, most of the operations involved locating the "playing cards" (i.e., High Value Personnel like ex-Iraqi military, Ba'ath party members, or government officials like Chemical Ali). A little later it was things like hitting militia hideouts, Al-Qaeda inspired cells, surveillance on likely targets, etc. In addition to the daily routine of house assaults there were a few team tasks with SIS ongoing simultaneously, with soldiers being pulled out from the squadron's strength. There were two SIS (MI6) compounds, one in the Green Zone and the other in the Palace in Basra, the latter of which headed up Operation Hathor in 2006. There were good relations between the SAS and Delta Force with a member of either team being exchanged.

By the time of the second Iraqi war, the Demarco became the main weapon of choice. However, Minimi machine guns, shotguns, and Glock or Sig pistols were also used as standard. For entry the demolition personnel would use strip charges, with Blade being the new explosive of choice.

Chapter 10

AFGHANISTAN

It is important for the reader to understand the basic nature of Afghanistan, which by the early twenty-first century was in a real mess. Afghanistan is an immense, landlocked country approximately the size of Texas with a population of around 24 million. It is a land of massive mountain ranges and remote valleys in the north and east, and near desert-like conditions on the plains to the south and west. Road and rail communications networks are minimal and in disrepair. However, prior to 1970, Afghanistan had been remarkably self-sufficient, especially in the production of food. Then in 1979 the Soviet invasion and contested occupation destroyed what political and economic structures were in place. Local tribal structure in the country remained strong but prevented the functioning of a strong central government, mainly because no one could agree.

The war against Russian occupation left an estimated 1.3 million Afghans dead or missing and created approximately 5.5 million refugees. While many countries condemned the Soviet Union's war, few did little to confront them. The CIA and other intelligence agencies covertly supplied arms and money to the Afghan mujahedeen (holy warriors), and even provided training in some highly sophisticated arms, such as the Stinger shoulder-fired anti-aircraft missile. In the end, the difficult terrain, huge losses, and the undeniable fortitude of the Afghans defeated the Soviet army, which departed in 1989. At this point all outside interest in Afghanistan declined and the country lapsed into near anarchy. The increase in tribal fighting added to Afghans' troubles until one group, the Taliban, established itself as the leader. The Taliban, which started in 1996, consisted of Muslim fundamentalists who sought to return the country to strict Islamic rule using whatever brutality was necessary in the process. By 2001, most of the country was under Taliban control except for some small areas held by Northern Alliance forces in the Panjshir Valley northeast of Kabul and a few scattered pockets of resistance in the northwest of the country.

When, on September 11, 2001, two hijacked airliners crashed into the twin towers of the World Trade Center in the United States, attitudes towards Afghanistan changed. More than three thousand US citizens and other nationals were killed as a result of the aggressive acts against America. The United States quickly uncovered evidence as to the identity of the perpetrators, and it very soon became clear that al-Qaeda were responsible for the attacks.

Taliban brutality is horrendous even amongst themselves.

While the bombings of American embassies in Nairobi, Kenya, and Dar-es-Salaam, Tanzania, on August 7, 1998, were clearly down to al-Qaeda, it was the direct attacks on the United States that started the war in Afghanistan.

IRAQ TO AFGHANISTAN

The 2007 migration from Iraq to Afghanistan signaled one of the largest ever deployments of British Special Forces; with two squadrons from 22 SAS, M Squadron SBS, elements of SRR and two companies of SFSG, the total force being close to five hundred personnel. The switchover of the SAS from Iraq to Afghanistan coincided with the replacement of General David McKiernan, the overall American commander of NATO's International Security Assistance

Force, with Lieutenant-General Stanley McChrystal, former commander of US Special Operations. McChrystal was staying with the men he loved to work with. McChrystal knew the SAS, and the SAS knew McChrystal. The effectiveness of the SAS in Afghanistan would be a reflection of the hard work done in Iraq.

Author's Note: The war in Afghanistan bled both the American and British armies severely, and the SAS also suffered. Special Forces recorded the highest number of deaths and casualties since the Second World War with twelve members killed and over eighty being severely wounded. Remember: the SAS is only two hundred strong.

While many terrorist acts were attributed to al-Qaeda, Osama bin Laden's involvement in the bombings of two American embassies was incontrovertibly proved. At the time it seemed as if al-Qaida could commit any atrocity it liked while hiding under the protection of the Taliban rule in Afghanistan—that was, until the attacks on America.

America fought back. However, armies take time to assemble and intelligence needed to be gathered. The Afghans had a ferocious reputation for fighting, especially in their own country. The first people to enter Afghanistan were a special unit of the CIA called the Special Activities Staff (SAS). These teams were staffed by retired members of Delta Force and SEAL Team 6, all well experienced field operators. They arrived on the ground in Afghanistan at the end of September 2001, in order to prepare for a major Special Forces operation.

A squadron of SAS arrived soon after and began operating in the mountainous country. Their first task was to concentrate on finding Osama bin Laden and Mullah Mohammed Omar, the Taliban leader. Later they would help train and lead the soldiers of the Northern Alliance in their push south towards Kabul.

It is also widely believed that the SAS took part in the incident at Qala-i-Jangi, where hundreds of Taliban soldiers were killed. On Saturday, November 24, some four hundred Taliban soldiers had surrendered to the Northern Alliance a few miles to the north of Mazar-i-Sharif, having fled the

American bombardment of Kunduz. The prisoners were then shipped by truck to a holding area in Qala-i-Jangi. This is a sprawling nineteenth-century prison fortress to the west of Mazar where the Northern Alliance warlord, Dostum, stabled his horses.

None of the Taliban soldiers were searched until they were inside the prison. At which time one of the prisoners produced a grenade and pulled the pin, killing himself and an Alliance commander called Nadir Ali. A second incident occurred later the same night when another prisoner killed himself and a Hazara senior commander, Saeed Asad. Despite these attacks, the Alliance guards did not seem too disturbed; neither did they arrange any extra security.

Scene from the Taliban Prison where SAS and US Forces put down a prison break.

Around 10 a.m. the next morning, two CIA officers entered the prison with the purpose of identifying any al-Qaeda members. The senior of these two Americans, Johnny Michael Spann, thirty-two, had operated in Afghanistan since the beginning of the war; his colleague was identified by the name of "Dave." They were taken by Alliance guards to an open area outside the cells where a group of Taliban prisoners had been assembled. As they were being interrogated, one prisoner leapt forward to grab at Spann's neck. Spann drew his pistol and shot the man dead. A fight then broke out resulting in the two Americans and the guards firing at the prisoners. One prisoner then managed to grab an AK-47 from an Alliance guard and opened fire. With this the Taliban fighters launched themselves at Spann, knocking him to the ground. Spann continued to fire his pistol, killing two more before he disappeared under the crush. Dave beat a hasty retreat as the Taliban overpowered the remaining Alliance guards, taking their weapons in the process. This group then ran to free the other prisoners before assaulting a nearby armory where they obtained AK-47s, grenades, and a mortar. A firefight then ensued with the Alliance soldiers holding the southeast quarter of the fort while the prisoners held the southwestern quarter. Some three hours after the incident had started, two vans and a pair of SAS Land Rovers arrived at the fortress gates. This force consisted of nine American Special Forces and six British SAS. There was a quick conference with the Alliance commander while one of the newly arrived Americans talked to Dave on the radio. Dave outlined the position; he was stuck with a German television crew and had run out of ammunition. He confirmed that the Taliban prisoners were armed and Mike Spann was missing in action (MIA). Dave was told to stay put while they dealt with the Taliban, assuring him that he would be rescued. While the fighting continued unabated, the Special Forces called for air cover. Shortly afterwards two American fighter planes arrived over the scene. At 4 p.m. the first missile hit, sending shock waves around the whole prison. The fighters continued to fire missiles for the rest of the afternoon, reducing much of the prison to rubble. Later that night, Dave and the German journalists managed to escape over the north wall.

By Monday morning, most of the Taliban had been killed or wounded and the Alliance had retaken much of the prison. Then disaster struck. At around 11 a.m., another air strike was called, and while the previous strikes had been close, this one was too close. The explosion killed a number of Alliance soldiers as well as wounding several Americans and members of the SAS. The firefight subsided as the casualties and walking wounded were loaded into makeshift transport, which sped off to the US base. In all, nine men were airlifted out. The next day, the Pentagon said that there had been no military deaths, but that five US service members had been seriously injured and had been evacuated to Landstuhl Regional Medical Center in Germany. Four SAS were also reported wounded, but the British Government never comment on their Special Forces.

Towards midnight an American AC-130 gunship arrived over Qala-i-Jangi. Having identified points of Taliban resistance, it proceeded to strafe the area. At some point one of its cannons hit an ammunition dump creating a massive explosion that could be heard 10 miles away.

By Tuesday morning there was only a handful of Taliban offering resistance and the Alliance started to move in and clear the prison. One prisoner who had escaped during the night was caught by local residents and hanged from a tree. The rest lay dead or dying in the hundreds, some still with their hands bound behind their backs. In one of the basements, five Taliban fighters were trapped alive. The Alliance soldiers threw in grenades before entering under a barrage of AK-47 fire. In the end, eighty-six filthy and hungry prisoners emerged; one of them was an American who had converted to Islam. Mike Spann's body was recovered, but only after a specialist team had been flown in to remove the booby trap that had been set under his body. There are two interpretations of the incident: some say the Taliban at Qala-i-Jangi had fought to the death, while others say they were massacred.

The SAS had been in Afghanistan for several weeks when they were ordered by the American Central Command in Florida to raid an opium storage plant to the southeast of Kandahar. The target location was said to hold some £50 million worth of pure opium and was heavily guarded. In addition to the opium, it was believed that, owing to the large number of guards (estimated at eighty-plus) on site, al-Qaeda might have some interesting intelligence documents at the location.

Unfortunately, maps of the area were virtually non-existent and the SAS were forced to rely on aerial photographs which had been taken by an unmanned US spy drone. The American command requested that the site be attacked almost immediately, leaving no time for a CTR. This would have enabled the SAS assaulting forces to better determine the strengths and weaknesses of the al-Qaeda position and refine their strategy accordingly. Timing also meant that the attack force would have to go in during daylight hours, abandoning the element of surprise and the use of their technically advanced night vision equipment.

SAS Land Rover. They used to be bright pink but are now more in keeping with desert sand. These men are on the range practicing with the Mark 19.

The SAS mustered two full squadrons—almost 120 men—the largest force it has fielded in a single operation for many years. The final assault was to rest on firepower. To this end, the SAS used some thirty Pink Panthers (110 Land Rovers especially made for desert warfare), all armed to capacity with ferocious weaponry. The Pinkies carried both front- and rear-mounted GPMGs (General Purpose Machine Guns), while the roll bar was mounted with either a Milan anti-tank rocket system, or Mk19 automatic grenade launcher. These were supported with big 2.5 ton ACMAC trucks acting as "mother" vehicles. Just prior to the assault, the American Air Force would carry out two precise air strikes. These would be close enough to destroy the drug warehouses, but leave the headquarters building intact.

The journey to the target started long before dawn; it was a long journey over dreadful, rocky terrain. As the area was hostile, the column was forced to stop at regular intervals and send out scouts. For this they used off-road bikes just as they had done in the Gulf war.

Dressed in light order, which means no rucksack, each SAS man carried his personal weapon which was either a Colt with an under-slung 203 grenade launcher or a Minime. During such an assault, the SAS prefer to forgo any body armor, depending instead on speed of movement. However, in this case both body armor and Kevlar helmets were used. Belt equipment consisted mainly of grenades, ammunition, food, water, and emergency equipment such as a SARBE beacon with which to call for help. By 10 a.m., the force was within two kilometers (a little over a mile) of the al-Qaeda camp but the going was slow owing to soft sand. This meant an on-the-spot change of plans.

Midday saw the start of the action as half the SAS lined up their Pinkies and started to lay down covering fire, ready for the other half to assault. The ground was not suited for a direct assault and the al-Qaeda rebels had chosen their defensive positions accordingly. By 1 p.m. a serious firefight had developed, but the assaulting squadron continued to push forward. The assault was helped at this stage by the arrival of US F-16s. The first pass caused massive explosions in and around the compound but as requested, didn't touch the building serving as the headquarters. The second air strike came close to wiping out the assaulting force. Conversely, as by this time the resulting drug-laden dust from the destroyed warehouses had thickened the air, nobody really cared very much.

Despite the horrendous fire put down by the SAS the al-Qaeda fighters refused to give ground, preferring to die. However, by 2:30 p.m., most resistance had been overwhelmed and the headquarters was raided for laptops, computers, papers, and maps. The ground was absolutely covered with dead al-Qaeda fighters and with those who were wounded. On the SAS side, several men had been hit, but thanks to the body armor and Kevlar helmets, most of the wounds were to the limbs. That said, several were extremely serious and required all of the skills of the SAS medics to keep them alive.

Author's Note: Even SAS soldiers talk, especially when they have had a few beers; this story in particular comes up several times, even though there is no precise proof that it is plausible. The stories depicted hand-to-hand combat and many other such actions, but the best one was the blowing up of one of the opium storage plants by the America aircraft. It would seem the entire surrounding area was covered in a dust cloud of opium that had everyone rolling around in tears.

Finally the word was sent to bug-out. Everyone jumped back in their vehicles and headed south. The first port of call was a rendezvous at a makeshift airfield where a C130 complete with doctors and medics evacuated the wounded, flying them directly back to Britain for treatment in the Center for Defense Medicine, Birmingham. The worst hurt of the four was hit in the stomach, arm, and the top of one leg. He was in a stable condition but was still faced with the prospect of losing his leg. Despite this, the mission had been accomplished without the loss of a single SAS soldier.

Princess Ann being trained in hostage rescue techniques at the old killing house in Hereford.

One thing the SAS are particularly good at is hostage rescue; for this reason they were selected to rescue a CIA operative who was being held captive in Kandahar by al- Qaeda. The building had been targeted when a Taliban vehicle had been spotted arriving and as such was put under observation. The agent was moved to a house surrounded by a high wall and protected by at least thirty Taliban fighters. According to the story, the American military were pondering the problem when an SAS officer happened to hear of the situation. "We can do it!" The surprised Americans were asked for a helicopter in order to perform the rescue and this was rapidly forthcoming.

It is rumored that about eight SAS soldiers then boarded the helicopter and flew directly to the building in question. Landing on the roof, half the team then proceeded to terminate any Taliban that showed their face, while the other four entered the building in order to carry out the rescue. They found the CIA agent strapped to a chair, suffering from a serious bout of physical interrogation. Again according to rumor, the SAS dispatched everyone in the room, grabbed the agent and made their way back to the roof. The area was fairly secure at this time as just about every Taliban within sight had been killed. The extraction went off without any one getting injured.

This story has never been validated, but shortly after, several members of the SAS were to visit the White House where they received medals. Additionally, the CIA, whose praises for the British SAS know no bounds, have sought to use them in other operations.

On January 17, 2002, Lieutenant General Cedric Delves took over the British seat at Central Command in Tampa, Florida, where American general Tommy Franks oversees the military effort against al-Qaeda in Afghanistan. Delves had commanded D Squadron during the Falklands War winning the DSO. His appointment indicated that Special Forces operations in Afghanistan would be on the increase. However, even with the SAS busy in Afghanistan, preparations were also underway for the second invasion of Iraq. On Thursday, March 20, 2003, at approximately 5:35 in the afternoon, the first salvo of missiles fell on Baghdad. After the Gulf War of 1991, when Saddam Hussein was left in change of Iraq, he failed to comply with UN resolutions to destroy weapons of mass destruction. The process of UN weapons inspections had come to an embarrassing end and America, together with the United Kingdom and other allies, decided to invade. It was not the war the world expected; the UN and NATO failed to support the Americans and took no part in the conflict. In the end, the task fell mainly on American and British soldiers.

While the SAS has always had a good working relationship with the American Special Forces, the past twelve years have seen a much stronger bond develop. In recent conflicts such as Afghanistan and Iraq, the Americans and the SAS have worked as one operational unit, their missions directed by the Department of Defense.

Those soldiers who do not live in Credenhill camp, live among the local community. Each morning all travel into the camp, pass the heavily armed security patrol and go about their daily business. Work for the average SAS soldier is learning, staying fit, planning. The SAS never rests. Every day, fifty of its members stand ready on the counter-terrorist team. They wait, and they train. Others kiss their wives, girlfriends, or partners good-bye and disappear for a time. This is the way of SAS life. They go and they return—but not all of them.

The year 2007 saw the SAS move from Iraq to Afghanistan which signaled one of the largest ever deployments of British Special Forces; with two squadrons from 22 SAS, M Squadron SBS, elements of SRR and two companies of SFSG, the total force being close to five hundred personnel. The switchover of the SAS from Iraq to Afghanistan coincided with the replacement of General David McKiernan, the overall American commander of NATO's International Security Assistance Force, by Lieutenant-General Stanley McChrystal, former commander of US Special Operations

McChrystal knew the SAS and the SAS knew McChrystal. The effectiveness of the SAS in Afghanistan has been as a result of the hard work done in Iraq during operation Task Force Black and Operation Task Force Knight. Cooperation between the American Special Forces and those of other nations has built a strong working bond and a set of engagement rules that work. This liaison is linked together through good intelligence; intelligence gained by surveillance and reconnaissance, which all lead to one goal, the execution. But this effectiveness comes at a price; the SAS are literally working themselves to death. As General McChrystal pointed out, in 2007 during A Squadron's six-month tour (180 days), they carried out 175 operations. McChrystal even took part in several of the SAS raids, all of which were at night, and became a familiar with the way the guys on the ground operated.

He was present on the night that the SAS mounted an operation to capture IED bombers in the area of Salman Pak just outside Baghdad. During the entry phase one of the Puma's went down killing two members of A Squadron, Sergeant John Battersby and Trooper Lee Fitzsimmons; twelve others were serious wounded. The rest of the assault party made an heroic rescue as the crashed helicopter rolled over on its side and burst into flames. When it was obvious that there was little more to be done, the survivors continued on to their objective and assaulted the house. McChrystal said later he was very humbled by the dedication the men had shown, knowing that the two men they had lost had been long-time friends and comrades.

IED bombs became the hallmark of the Afghan War, many being triggered by mobile phones. They were a serious danger to all the troops serving in Afghanistan.

HOSTAGE RESCUE IN AFGHANISTAN, 2012

On May 22, 2012, four women working for a medical charity in Afghanistan were seized, among them a British woman named Helen Johnston. They had been taken while traveling on horseback from Yafta to the Yavan districts of Badakshan. The kidnappers, who were known to have close links to the Taliban, released a video in which they demanded the release of some Taliban prisoners and £6 million in ransom. The hostages were believed to be alive and well, but worryingly the kidnappers were known to have been in contact with a small pocket of Taliban fighters.

Over the next few days, SAS commanders, intelligence officers, and members of the Afghan National Defense Directorate watched the kidnappers' activities on real-time video transmitted via satellite from the Predator as they prepared the rescue plan within the headquarters of the Joint Special Forces Group in Kabul.

The four women were being held in a cave deep inside the thick Koh-e-Laram Forest of north western Badakshan, a remote region close to the Tajikistan border. The hostages' exact location was isolated using mobile phone interception technology and Predator drones were dispatched to silently keep the location under twenty-four-hour surveillance.

Nonetheless, analysts had received intelligence that the hostages had been split into two groups and were being held in separate caves. Concern for the women's safety increased when a member of the Taliban was overheard in an intercepted mobile phone conversation pressuring the kidnappers to put on a "show of intent," which meant the possible death of one or more of the women.

Both the United States and United Kingdom agreed to mount a joint raid using a SEAL team and members of the SAS. The combined team was moved to a forward operating base in Badakhshan from where they were flown some two miles from the cave complex where the hostages were being held. The rugged mountain terrain was hard going but the thick forest also supplied some protective cover for the rescue team. Back in Kabul, the commanders watched their progress live as it was streamed from the helmet cameras worn by the soldiers and from the Predator drone, both which were connected via satellite. At the front of everyone's mind was the death in 2010 of aid worker Linda Norgrove, who had been killed in a similar rescue attempt in North Eastern Afghanistan.

By early evening they were in position to assault, albeit with stealth. As they engaged the kidnappers, a firefight alerted both caves. Within minutes the US SEALs reported their cave did not contain any hostages—only to be told that the SAS had been successful and that the four hostages had been released. Miss Johnston, a committed Christian, along with a Kenyan colleague, and two Afghan women who worked for the same aid agency, were said to be physically well after their ordeal—eleven kidnappers were killed during the assault which took just three minutes.

Ironically, while commanders from both America and Britain congratulated the Navy SEALs and SAS on a job well done, another British soldier from the 1st Battalion Royal Welsh Regiment was killed in a firefight while on patrol in Helmand province.

Taliban Cave complex found by US Navy SEALs.

SUMMARY

Afghanistan taught Special Forces a hard lesson; the Afghan fighters are tough, clever, and happy to die for their cause. Given the massive presence of American, British, and a whole host of other countries that participated in the war, one would have thought that for once they would prevail. Nevertheless as this book goes to print, the war in Afghanistan is coming to an end, but I fear the war has not been won.

Yet many lessons were learned, but at a high cost. One of the most horrendous devices to come to the forefront was the improvised explosive device (IED). This simple explosive device made from anything available; old artillery shells, to homemade explosives torn at the heart of the allied armies. It killed and removed thousands of limbs, caused patrol tactics to be changed and scared the living daylights out of the soldiers. Despite the millions of dollars spent looking for a solution, nothing acceptable was ever found.

Notwithstanding, Special Forces of America and Britain worked relentlessly, raiding caves and houses alike, seriously depleting the Taliban leadership. Within this struggle a much stronger bond evolved between the British and Americans.

BIBLIOGRAPHY

Barber, Noel, *The War of The Running Dogs: The Malayan Emergency, 1948–1960*, Collins, 1971.

Benyon-Tinker, W. E., *Dust Upon The Sea*, Hodder & Stoughton, 1947.

Billiere, General Sir Peter de la, *Looking For Trouble: SAS to Gulf Command*, BCA, 1994.

Bonds, Ray (edited by), *The Vietnam War: The Illustrated History of The Conflict in South-East Asia*, Salamander Books, 1979.

Bradford, Roy & Martin Dillon, *Rogue Warrior of The SAS*, John Murray, 1987.

Cole, Barbara, *The Elite: The Story of The Rhodesian Special Air Service*, Three Knights Publishing, 1984.

Cole, Barbara, *The Elite Pictorial: Rhodesian Special Air Service*, Three Knights Publishing, 1986.

Cowles, Virginia, *The Phantom Major: The Story of David Stirling and The SAS Regiment*, Fontana Books, 1958.

Courtney G. B., *SBS in World War Two*, Grafton Books, 1985

Davies, Barry, *The Complete Encyclopedia of the SAS*, Virgin Books, 1998.

Davies, Barry, *SAS Rescue*, Sidgwick & Jackson, 1996.

Draper & Challenor, *Tanky Challenor: SAS and the Met*, Leo Cooper, 1990.

Dickens, Peter, *SAS: Secret War in South-East Asia*, Greenhill Books, 1991.

Farran, Roy. Winged Dagger. Arms & Armour Press 1986.

Farran, Roy. Operation Tombola. Arms & Armour Press 1986.

Generous, Kevin, *Vietnam: The Secret War*, Hamlyn Bison, 1985.

Geraghty, Tony, *Who dares Wins: The Special Air Service, 1950 to The Falklands*, Arms & Armour Press, 1983.

Harclerode, Peter, *PARA!: Fifty Years of The Parachute Regiment*, Arms & Armour Press, 1992.

Harrison, Derrick, *These Men Are Dangerous: The Early Years of The SAS*, Blandford Press, 1988.

Hoe, Alan, *David Stirling: The Authorized Biography of The Creator of The SAS*, Little, Brown & Co., 1992.

Hoe, Alan & Eric Morris, *Re-enter the SAS: The Special Air Service and the Malayan Emergency*, Leo Cooper, 1994.

Horner, David, *Phantoms of The Jungle: A History of The Australian Special Air Service*, Allen & Unwin, 1989.

James, Harold & Denis Sheil-Small, *The Undeclared War*, New English Library, 1973.

Kemp, Anthony, *The SAS: The Savage Wars of Peace 1947 to the Present*, John Murray, 1994.

Ladd, James, *SAS Operations*, Robert Hale, 1986.

Ladd, James, *SBS: The Invisible Raiders*, Arms & Armour Press, 1983.

Ladd, James & Keith Melton, *Clandestine Warfare: Weapons And Equipment of the SOE and OSS*, Blandford Press, 1988.

Langley, Mike, Anders Lassen VC, *MC of the SAS*, New English Library, 1988.

Lassen, Suzanne, Anders Lassen, *The Story of a Dane*, Frederick Muller, 1965.

Lodwick, John, *The Filibusters*, Methuen & Co., 1947.

Lorain, Pierre, *Secret Warfare: The Arms And Techniques of The Resistance*, Orbis Publishing, 1983.

Malone, M. J., *SAS: A Pictorial History of The Australian Special Air Service, 1957–97*, Access Press, 1997.

Parker, John, *SBS: The Inside Story of The Special Boat Service*, Headline, 1997.

Pitt, Barrie, *Special Boat Squadron: The Story of The SBS in The Mediterranean*, Century Publishing, 1983.

Pocock, Tom, *Fighting General: The Public & Private Campaigns of General Sir Walter Walker*, Collins, 1973.

Ramsay, Jack, *The Soldiers Story*, Macmillan, 1996.

Seligman, Adrian. *War in The Islands: Undercover Operations in The Aegean, 1942–44,*. Alan Sutton Publishing, 1996.